ABOUT LIFE

About Life

Concepts in Modern Biology

by

PAUL S. AGUTTER
Theoretical and Cell Biology Consultancy,
Glossop, Derbyshire, U.K.

and

DENYS N. WHEATLEY
BioMedES, Inverurie,
Aberdeenshire, U.K.

 Springer

A C.I.P. Catalogue record for this book is available from the Library of Congress.

ISBN-13 978-90-481-7363-1
ISBN-10 1-4020-5418-1 (e-book)
ISBN-13 978-1-4020-5418-1 (e-book)

Published by Springer,
P.O. Box 17, 3300 AA Dordrecht, The Netherlands.

www.springer.com

Printed on acid-free paper

TABLE OF CONTENTS

PREFACE

Thanks to the popular media, and to books by Dawkins, Fortey, Gould, Margulis and other writers, people are informed about many aspects of biology. Everyone seems to know a little about evolution, for example, and about DNA and the possibilities (good and bad) afforded by research in molecular genetics. Most people know some of the arguments for and against the likelihood of life on other planets. And so on. We are glad that these pieces of information have become so widely available. However, we do not assume any particular knowledge (other than the most basic) in this book. Our aim is to address *general* questions rather than specific issues. We want to enable our readers to join their disparate pieces of knowledge about biology together.

The most basic of these general questions – and perhaps the most difficult – can be expressed in beguilingly simple words: "What is life"? What does modern biology tell us about the essential differences between living organisms and the inanimate world? An attempt to answer this question takes us on a journey through almost the whole of contemporary cell and molecular biology, which occupies the first half of the book. The journey is worth the effort. The provisional answer we attain provides a coherent, unifying context in which we can discuss evolution, the origin of life, extraterrestrial life, the meaning of "intelligence", the evolution of the human brain and the nature of mind. In other words, it enables us – as we said - to help our readers to join their disparate pieces of information together.

Although we assume virtually no knowledge of biology and use non-technical language as far as possible, we cannot avoid using some technical terms. These will be unfamiliar to many readers, so we have added a glossary and pronunciation guide after the final chapter.

We intend this book to be the first volume of a trilogy. In the second volume we plan to explore what science is, and why scientific thinking originated and flourished in western society. We want to investigate the ways in which biology resembles other sciences and the ways in which it differs from them. In the third book, we hope to explore the most controversial topics associated with biology today: patenting of human genes, cloning, genetic modification of crops, the obliteration of habitats, the extinction of species, and so on. This first volume is a prelude to these future projects.

We are grateful to many colleagues for discussions and advice during the several years of gestation of this book, and to the Carnegie Trust for a grant to support the project. All the illustrations were prepared by Dr Ruth Campbell, whose diligence in this work we gratefully acknowledge. Some of the illustrations are reproduced with permission from published sources: Fig. 2.1 from Goodsell (1991) "Inside the living cell," *Trends Biochem. Sci.* **16**, 206-210; Figs. 3.1(a) and 7.1 (b) from Mayer, Wheatley and Hoppert (2006) in *Water and the Cell*, chapter 12, Springer, Dordrecht; Figs. 5.3 and 13.5 from de Robertis and de Robertis (1980) *Cell and Molecular Biology*, 7th edition, Saunders, Philadelphia; Fig 6.5 from http://personalpages.umist.ac.uk/staff/goughlecture/the-cell/diffdev3/haemo.jpg; Fig. 8.3 from Wheatley (1982) *The Centriole: a Central Enigma of Cell Biology*, North Holland Biomedical Press; Fig. 10.2 from Hogben (1958) *Science and the Citizen*, George Allen and Unwin; Fig. 10.3 from <http://www.ug.edu.au/school> science lessons /3.0; and Fig. 12.2 from <http://steve:gb.com/images/science/hydrothermal.jpg>.

While we have done our best to distil the basic concepts that guide biology today, informed readers are likely to consider parts of the text to be in need of revision or correction. We shall be glad of critical feedback. Science is a collective activity, and we are part of the collective.

PSA
DNW

Chapter 1

INTRODUCTION

On a fine day in late spring or early summer, preferably around sunrise or sunset, go to a patch of uncultivated or wooded land as far as possible from people and traffic. Find a comfortable place where you can remain quiet and still for half an hour. Wait, watch and listen. For a while you hear only the sounds of insects, the alarm calls of small birds and the breeze among foliage; nothing moves except leaves and clouds. But after ten or fifteen minutes there is a transformation. Birds settle and feed. Shiny beetles sidle down tree trunks and over the ground. Furry bodies dart to and fro. The world around you has come alive.

Such experiences bring us into contact with other species and seem to satisfy a deep human hunger. "Communing with Nature" is sometimes said to refresh the spirit. The sights and sounds and smells of non-human life in its natural setting arouse our curiosity. They fascinate and enchant. They are the source of much poetry, music and visual art - and of science.

Science - in this case the science of life, biology – has its roots in curiosity. What we see raises questions. These might be simple questions, such as the names of the trees and the shiny beetles and the owners of the furry bodies. Or they might be more complicated ones, such as how birds and flowers are made, how they do the things they do, why they do them; and why they exist at all. *Science is a way of framing such questions and trying to answer them.* It is not the only way, but it is a very informative and productive one. It works by considering things in themselves, taking no account of whether they are beautiful or ugly or good or bad. The nature and origins of science and its effects on the world are topics for a different book. For present purposes a simple definition will suffice: *science is a way of satisfying our curiosity by formulating questions about what we observe and answering them dispassionately* – that is, without making value judgements.

You might ask how "communing with Nature" can still enchant a person - a scientist - who devotes his or her working life to dispassionate analytical inquiry. Surely, when curiosity is satisfied, wonder is lost? In fact, for most scientists, the opposite holds. Understanding the techniques of counterpoint and sonata form can enhance our appreciation of Bach fugues and Beethoven symphonies. Analysis of literary styles can help us to relish the subtle ways in which Henry James or Charles Dickens convey character and tension and a sense of place. In much the same way, the fruits of scientific inquiry increase both our understanding of the natural world and our wonder at its workings. Framing and answering questions does not destroy our pleasure in what we see around us; quite the contrary. Knowledge (especially the acquisition of knowledge) is pleasurable in itself, and it augments other pleasures.

However, research scientists ask different *kinds* of questions from other people. Two individuals who witness the same burgeoning of life during a spring sunrise might experience similar feelings of wonder and excitement. But if one of them is a practising biologist and the other is not, their curiosity will take different forms. The non-specialist might ask why certain insects visit primroses but not wood anemones, or how the shiny beetle manages to feed on the unappetising trunk of the oak tree; or how swallows, swooping from the bright sunlit air into the windowless barn, adapt so quickly to the sudden darkness that they unerringly find their nestlings and never collide with beams or walls. The specialist, the scientist, might be able to answer such questions; if not, then answers will surely be found among the wealth of available wildlife documentaries, books and magazine articles. But personally, he or she will be interested in different matters: the exact mechanism, say, whereby the primrose flower synthesises its chemoattractant, and why insects of one species but not others respond to it; or precisely what place the shiny beetle has in the ecology of mixed woodland. For both individuals, the pleasure of questioning and answering enhances the immediate sensory experience. But the biologist's pleasure in knowledge is difficult to share, except with those who have the same specialist background. There is a comprehension barrier, which we need to try to cross so that scientific knowledge becomes more generally accessible.

Popular science books, television and radio documentaries, science articles in newspapers – all these have gone a long way towards overcoming this barrier. Nevertheless scientists still tend to feel, and to be, misunderstood. This is apparent in their reactions to the most general, basic-seeming questions, the sorts of questions that a child might ask. Scientists tend to consider such questions unanswerable: too vague, too resistant to accepted technical vocabulary, too remote from the rigorous demands of ongoing research; in a word, too hard. For instance, when after a quarter of an hour's stillness in a chosen rural spot you have merged into the landscape and the world has "come alive" around you, what exactly does that phrase mean? Of course it "means" that local animal life has come out of hiding and revealed itself, but *what do the words "alive" and "life" really denote*? What fundamental properties do the primrose, the oak tree, the beetle, the swallow and the darting weasel have in common that distinguish them from the soil and rock beneath them, the air around them, the clouds above them, or the sunlight on which they all ultimately depend?

We can broaden this question. The bodies of the oak, the swallow and every other plant and animal are swarming with microscopic inhabitants such as bacteria. So is the soil itself. There are probably more bacteria in a handful of soil than there are leaves in the entire wood. What properties do these minute living things, scraps of matter that cannot be seen without a powerful microscope, share with the primrose and the beetle and the weasel but not with anything inanimate? *What is "life"*? Many have asked this question. It is the main topic of this book.

Thanks to a number of excellent popular scientific publications, most people nowadays might answer "What is life?" by saying "DNA". All living things contain DNA, but no inanimate ones do. DNA is the material of the coded instructions - the *genome* - for making and maintaining an organism. Cracking the code, unravelling the genome sequence, helps us to understand everything there is to know about that organism. The non-living world has no genome, no coded instructions. That is the difference between the living and the inanimate. For very good reasons, this answer has become deeply entrenched in modern thought: life is DNA. The double helix has become a major cultural icon. The complete sequencing of genomes (not least the human genome) has been hailed as one of the greatest achievements of human history.

However, without belittling this achievement or doubting that DNA is indeed basic to life on Earth, we can challenge the answer. Indeed, *is* it an answer? DNA itself is not living. Pure DNA in a test tube does not behave like anything alive; in fact, it does not behave at all. A freshly fallen leaf contains just the same DNA as it did before it fell, but it is no longer alive. Moreover, the fallen leaf still contains the materials – the proteins and their products - that the genome instructed it to make. So non-living things such as test tubes, and once-living things such as dead leaves, can contain DNA *and* the substances that DNA codes for. Yet they are not alive.

There are other objections, too. For instance, there might be entities on planets far across the galaxy that we would (if we ever saw them) describe as "living" because they shared certain characteristics with terrestrial organisms. Suppose we could analyse one of these hypothetical entities, and suppose we found that it contained no DNA. Would we then declare: "Our mistake. Despite appearances, these entities aren't living after all"? Surely not. So we are back where we started. We might accept that DNA is fundamental to life on Earth (whatever we mean by "fundamental" and "life"), but neither DNA nor the materials it encodes are sufficient to define the living state. The question "What is life?" remains open.

Many biologists are impatient with the question. They point to past attempts to distinguish the living from the non-living (traditionally, organisms are said to *eat, breathe, excrete, grow, move, respond to stimuli* and *reproduce*) and tell us, quite rightly, that all such attempts have proved inadequate. The reason why they have proved inadequate is simple. "Eating" involves wildly different processes in, say, oak trees and weasels. Weasels "move" in ways that oaks do not. And so on. Any definitions of "eating" and "moving" that are broad enough to encompass such a range of meanings would be useless. They would apply to many non-living things as well as living ones; and however broad we made our definitions, there would probably still be living things to which they would *not* apply. The quest for a clear distinction between living and non-living has always been vain, say the sceptics, so it is a waste of time to consider the question further.

This attitude is understandable but it unsatisfactory. If biology is the study of life and we cannot define life, then we cannot define what biology is about. This elementary logic ought to make the sceptics uncomfortable. Also, if we cannot define life, what do we mean by the "origin of life"? The origin of what? Similar problems abound. One more example: an established tenet of biology is that *the cell is the fundamental unit of life*; in other words, every organism comprises one or more cells. (We shall start to explore what we mean by a "cell" in the next chapter.) But if we cannot define life, of what is the cell the fundamental unit?

Other biologists take a different view, less sceptical but not very helpful. The living, they say, can be distinguished from the non-living by our detailed knowledge of the workings of organisms, knowledge that we have acquired through centuries of research world-wide. In principle, this view is unexceptionable. Any definition or characterisation of the living state *must* be based on what we have learned through the progress of science. But the amount of published biological data is colossal. Consider cell biology alone. The workings of some types of cell, such as the intestinal bacterium *Escherichia coli* or a rat liver cell, are known in mind-numbing - though not yet exhaustive - detail. The existing mass of information about such cells is far too unwieldy to provide a comprehensible distinction between the living and the non-living. And what essential facts might lurk among the details we have not yet discovered? Moreover, though all living cells share many features, each type of cell is also distinctive; and the common features might not suffice to identify an object as "living". So although a general account of the living state must be firmly based on what we know about particular living cells, this approach to answering "What is life?" is impractical if we take it literally.

In this book we shall construct a provisional, somewhat abstract answer to the question "What is life?" by generalising from these masses of information. We shall express this answer in non-technical terms as far as possible. We believe that our answer is interesting enough to publish, but it is not written on tablets of stone. It will probably be challenged by other biologists; indeed, we hope it will. Science - like the communication of science - progresses by trying out ideas, finding flaws in them, and trying again. If no ideas are put forward there is nothing in which to find flaws and therefore no progress. So although we are prepared to defend our provisional answer, we want it to be a target for rational criticism. Rational criticism will lead to better answers.

The words *alive* and *life* define the main theme of this book, but we shall also look at some related issues. Some of these issues, such as the origin of life and the existence of extraterrestrial life, have received much attention from other authors. Our contribution, a small one, is to reconsider them in the light of our general "definition" (or rather *characterisation*) of the living state. Inevitably we shall discuss evolution - it is impossible to write a book about biology without mentioning biology's central theory - but again we shall take advantage of the excellent popular treatments of this subject that are already in print.

One question recurrently asked about extraterrestrial life is whether it might be "intelligent" in the sense that our species is intelligent. To consider this question, we shall briefly discuss the nature and evolution of the organ of human intellect, the brain. Once more we shall take advantage of popular accounts, and of the revolutionary progress made in neurobiology during the last two decades of the twentieth century; but we shall suggest a new perspective on the topic.

To summarise: we begin the book by focusing on the "fundamental unit of life", the cell, and we spend the first few chapters developing our characterisation of the living state. In chapter 11 we turn to evolution, and in the remainder of the book we consider the origin of life, the evolution of "intelligence" and the question of extraterrestrial life.

Our aim is to share ideas equally with fellow-biologists and non-specialists. We invite all our readers to challenge the central idea in this book, the fundamental difference between the living and the non-living, and to improve on it. Any reasonable attempt to answer "What is life?" will help to develop more coherent views about the origin and evolution of life on Earth, the nature and evolution of intelligence, the possibilities for extraterrestrial life, and other big topics.

It is enjoyable to debate these topics, so this seems a worthwhile aim in itself. But there is another point: to have clear ideas about such broad issues enhances the wonder and pleasure that we gain from contemplating the world around us. In consequence, our thoughts and reflections when we "commune with Nature" at sunset will continue long after the stars come out.

Chapter 2

INGREDIENTS OF THE SIMPLEST CELLS
Prokaryotes and the sizes of their contents

Cells are small. To see them you need a microscope, and to see their contents in detail you need an electron microscope. Objects so minute that they cannot be seen with the naked eye are - by definition - remote from everyday experience. This makes it hard to grasp the *scale* of cells and their contents. And without a grasp of scale it is impossible to acquire a clear mental picture of a cell.

In this chapter and the following one we shall describe large-scale models of cells that can be made from ordinary household materials. These models use the familiar to represent the unfamiliar. We urge our readers to *make* them. They are very simple, and entertaining to build if two or three people work together on them. Seeing and touching the models will create more vivid and memorable pictures than simply reading our instructions and comments. Building them will not reveal how cells work; we shall explore that in later chapters. But it will familiarise you with the main components of cells, and it will illustrate the relationships among these components and indicate their relative sizes. The relative sizes will prove surprising.

Before we begin on the models we must introduce two technical terms that might not be familiar to everyone. Terrestrial organisms are of two kinds: *prokaryotes* and *eukaryotes*. Prokaryotes are tiny one-celled organisms such as bacteria that do not contain a separate nucleus. "Prokaryote" is derived from Greek roots meaning "before the kernel (nucleus)". Eukaryotes are organisms consisting of one or more cells, each of which *does* have a separate nucleus containing the bulk of the DNA. "Eukaryote" comes from the Greek for "well-formed kernel (nucleus)". Single-celled organisms such as yeasts and amoebae are eukaryotes. So are all multicellular organisms: all fungi, all plants from mosses and seaweeds to primroses and oak trees, and all animals from sponges and worms to beetles and swallows and humans. (Most scientific terms come from Greek and

Latin, or occasionally Arabic, roots. This is because, until the early 20th century, science was the pursuit of gentlemen who were educated in the Classics, and much of our knowledge has Classical and Arabic foundations. Words that are not in common use and are employed only for special technical purposes have a great advantage: their meanings remain stable and unambiguous. For science students, the drawback of such words is that they have to be learned.)

Despite appearances, which are misleading because we can only see multicellular organisms - and not even all of those - with our unaided eyes, the world's prokaryotes greatly outnumber the eukaryotes. Also, prokaryotes are far more venerable: the earliest prokaryotes lived on Earth twice or three times as long ago as the most ancient eukaryote. Bacteria have a bad press because, for historical reasons, we associate them with infectious diseases. However, very few bacteria cause disease. The overwhelming majority are not only harmless, but in some cases essential for other forms of life. For example, if it were not for the bacteria that make atmospheric nitrogen available to plants, plants would not exist – and as a result, neither would any animals, including ourselves.

Let us return to the matter of cell size. A metre ruler is divided into a thousand parts – millimetres. Everyone knows that; we can *see* a metre ruler and its millimetre divisions. But try to imagine a *millimetre* ruler divided into a thousand parts. Each part would be a thousandth of a millimetre; that is, a millionth of a metre, or *micrometre*. ("Millimetre" is abbreviated to "mm". "Micrometre" is abbreviated to "μm". The Greek letter *mu*, μ, is the usual way of indicating "a millionth of".) This imaginary ruler is almost impossible to picture, but to measure cells we would need only a small portion of it. A typical prokaryote is just one or two micrometres long. Eukaryotic cells vary in size (for example, plant cells are usually bigger than animal cells), but a cell in your liver – to take an example at random – might be some fifteen or twenty micrometres across. A small eukaryotic cell is around ten times greater in linear dimensions (that is, ten times longer, wider and taller) than a prokaryotic cell. This means it is around a thousand times greater in volume. (Picture two cubes, one with one-centimetre edges and the other with ten-centimetre edges. The second cube has a thousand times greater volume than the first - one litre compared to one cubic centimetre – but the difference in linear dimensions is tenfold.) In other words, eukaryotic cells are much bigger than prokaryotic ones. They also have more complicated structures. Therefore, we are going to describe two different "household" models, one for each main type of cell. In this chapter we shall describe a matchbox-sized model for a prokaryote. In the next chapter we shall describe a cardboard carton-sized one for a eukaryotic cell.

A matchbox model of a prokaryote: the advantage of being small
An ordinary matchbox measures roughly 2 inches by 1 inch by 1 inch (5 cm x 2.5 cm x 2.5 cm). Prokaryotic cells are not exactly rectangular, as matchboxes usually are, but they have similar proportions. To make a matchbox model of a prokaryote, let one inch (2.5 centimetres) of matchbox represent one micrometre of cell. So the matchbox corresponds to a prokaryote of "typical" size, 2 μm x 1 μm x 1 μm.

We have magnified the cell to the size of a matchbox, so we must magnify its contents in proportion. First, we need to consider the DNA. Prokaryotic DNA is circular; we can represent it by cutting a piece of thread of suitable length and knotting the ends together. A real bacterial DNA is about one third to one half of a *millimetre* long (300-500 μm)[1], so the matchbox model will need to contain a thread that is between 25 and 42 *feet* (7.7 and 13 metres) long. When you cut a thread of this length and knot the ends together, the result is a tangled circle. When you push this circle into the matchbox it becomes even more tangled. Unless you have used very fine thread the circle will have overfilled the matchbox – a problem, since we have many more items to add to the model; DNA is only one of the cell's many constituents.

Although the model is far from complete it has already demonstrated some important points. First, it has shown that DNA is an extremely long molecule, hundreds of times longer than the cell that contains it. Second, DNA must also be a very thin molecule, or it would not fit into the cell however hard you pushed it. Third, when DNA is packed into a cell, it is twisted and folded into a shape undreamed of by the most mischievous kitten among knitting wool. Making the matchbox-and-thread model brings these points home convincingly.

Cell functions are not topics for this chapter, but most people know that DNA is the material of the genes. A gene is a segment of DNA. In the matchbox model, an average-sized gene is represented by about one centimetre of the 'DNA' thread. For the time being we shall assume that each gene codes for one cell protein. (This assumption is not exactly true but it will suffice until chapter 11.) To make the protein corresponding to one gene, i.e. the protein *encoded* in that gene, the cell needs the right equipment. This equipment includes various kinds of *RNA*; molecules

---

[1] Bacterial DNA is roughly one million bases long, a "base" being a single unit (letter) of the coded information that the molecule contains. In the commonest double-helical form of DNA, one base occupies a length of 0.34 nanometers, a nanometre being a thousand-millionth of a metre (i.e. a thousandth of a micrometre or a millionth of a millimetre). One million bases at 0.34 nanometres per base comes to 0.34 millimetres (340 micrometres) in total.

similar to DNA but much shorter and less stable. One sort of RNA, known as "messenger", is a copy or imprint of a gene or a small group of genes.

Think of the DNA as a library of master documents, none of which can be removed from the library but any of which can be photocopied. Each document is a gene, a coded instruction for making a particular protein. The messenger RNA molecules are the photocopies; they *can* be taken out of the library. Each messenger photocopy is fed into the *ribosomes*, remarkable machines that scan the photocopied document, translate its instructions and make the protein encoded in the gene. Thanks to this system, the instructions in a gene can be used for manufacturing thousands of copies of the same protein. Proteins are responsible for all the structures and activities of the cell: holding the cell together, sensing and responding to the environment, taking in nutrients and metabolising them, controlling the energy supply, manufacturing other cell constituents, copying DNA, making RNA, using ribosomes, and so on. The *proteins*, not the DNA that codes for them, are largely responsible for the "living state".

RNA molecules, ribosomes and proteins all need to be represented in the matchbox model. At any moment a prokaryote contains roughly as much RNA as it does DNA. So cut another ten metres or so of thread and put that into the matchbox. (For authenticity, you should snip this second piece of thread into 1-10 centimetre segments. This would represent the RNA molecules more realistically. However, repeated snipping is tedious and adds little to the point of the exercise.) A rounded teaspoonful of lentils represents the ribosomes. The cell's proteins can be represented by a rounded teaspoonful of sugar. If a prokaryote is magnified to the size of a matchbox, each protein molecule it contains is, on average, about the size of a sugar grain, and each ribosome is about the size of a lentil.

These proportions might seem hard to believe. Many biologists can remember feeling incredulous about them on first encounter. (One of the present authors recalls checking the calculations six times, sure there must be a mistake somewhere.) The circular 'DNA' thread in the matchbox model represents about 1000 genes, so the average gene corresponds to roughly one centimetre of thread. On the same scale, the protein encoded by the gene corresponds to a grain of sugar. Gene is to protein as a centimetre of thread is to a sugar grain. Yet that single grain of sugar, the protein molecule, is the whole *point* of the gene, because the proteins are responsible for virtually all the cell's structures and activities.

We have dealt now with most of the contents of a prokaryote: the huge circular DNA molecule, the many shorter RNA molecules copied from the DNA, the proteins that are necessary for the cell's structure and all its activities, and the ribosomes for making the proteins. There are tiny nutrient molecules as well, and the cell could not function without them, but *in toto* they occupy relatively little space. Some prokaryotes contain storage granules (food reserves), so add half a dozen dried peas to the matchbox to represent these. And of course there is water - about 20 ml in the matchbox model - but we would not recommend actually adding it; the results would be messy. Just imagine that the remaining space in the model is water not air, and ask yourself how the matchbox can accommodate 20 ml of it, considering the space occupied by the other contents.

Then close the box.

A question will strike you immediately. Why is everything packed so tightly? Why is the cell not bigger? If you have so much luggage to pack why not use a suitcase? The accepted answer is as follows. If you reproduce in the simplest possible way, that is, by duplicating all your contents and then splitting into two pieces so that one copy of everything goes into each half, it is an advantage to be small. The smaller you are, the less of you there is to duplicate, so the less time and energy it takes. Therefore, by being as small as an organism can be, prokaryotes maximise their reproductive rates. So their populations grow at the greatest possible speed - until they run out of nutrients.

Producing the maximum number of descendants is a basic biological "drive". The aim is the long-term survival of the genes. Explosive population growth helps to ensure this outcome. Because they are as small as possible and therefore reproduce as quickly as possible, bacteria can transmit their genes to large populations of descendants in the shortest possible time. At maximum growth rate, a bacterial cell may divide every twenty minutes. If there is one cell at time zero, there are two after twenty minutes, four after forty minutes, eight after an hour, sixty-four after two hours, and so on. Given an inexhaustible nutrient supply, there would be about 4,722,366,483,000,000,000,000 cells, 14,000 tonnes of solid bacteria, after a day's growth from a single cell. Of course the nutrient supply would run out long before that number was reached, but bacteria do proliferate very quickly under optimal conditions. It gives their genes the best long-term chance of survival.

The prokaryotic cell surface: the drawback of being small
The matchbox model has served its two main purposes: (1) to give a clear impression of the relative sizes of cell, DNA molecule, protein molecules and ribosomes, and (2) to show that a prokaryotic cell is about as small as it can possibly be - it is very tightly packed. The model has also led to a discussion of the relationship between cell size, population growth, and the biological imperative to transmit genes to future generations. (More about this "drive" later.) Like any model, however, it has limitations. We have mentioned two of these already: prokaryotes do not have sharp corners and

edges like a matchbox; and the model is static - it does not represent any of the myriad activities, including reproduction, in which a cell engages. There is another limitation as well, rather an important one: the model misleads us about the nature of the cell surface.

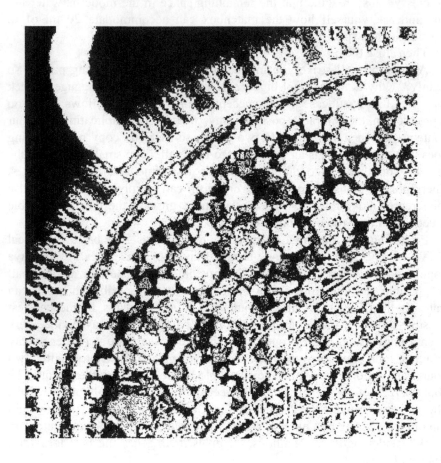

Fig. 2-1: A bacterial cell surface.

The cardboard of the box can be taken to represent the tough protective coat surrounding most prokaryotes, the *cell wall*. Inside the cell wall, however, there is a very thin continuous coat, the *cell membrane*[2], which in

---

[2] Some prokaryotes (Gram negative bacteria, for instance) have an extra membrane outside the cell wall, but we shall not develop this point here.

our matchbox model would have to be represented by an unbroken seal of polythene or cellophane about a quarter of a millimetre thick. Despite its unimpressive appearance, the membrane is one of the most active and versatile parts of the cell. It jobs include:-

- Separating the cell interior from the outside world, preventing inadvertent mixing.
- Controlling the flow of materials into and out of the cell.
- Enabling items of food in the outside world to be digested, and importing the products of digestion to the cell.
- Detecting stimuli (the whereabouts of food, the location of danger, etc) and initiating appropriate responses.
- Playing a central role in making energy available to the cell. In some prokaryotes, it traps energy from the outside world and changes this energy to a form that the cell can use. (Cyanobacteria, for example, trap the energy of sunlight.)
- Housing the equipment for manufacturing many of the cell's constituents, including the external wall.
- Initiating and controlling the duplication of the DNA - the essential step before cell division (reproduction).

The first two of these functions are clearly linked. The surface membrane must be a barrier but not an impermeable one; it has to be selective. It must enhance the entry of materials that the cell needs and the exit of waste products, but it must be a barrier to everything else. Designing a structure with such exacting properties would be an engineer's nightmare, particularly if the barrier had to be no more than ten nanometres (one hundred-millionth of a metre) thick, the usual thickness of biological membranes. Yet life on Earth produced this structure thousands of millions of years ago.

Because of the membrane's remarkable range of functions, being small is in some ways a *disadvantage* for prokaryotes. The smaller the cell, the smaller the area of the cell membrane. The smaller the membrane area, the less equipment can be fitted into it. The fewer the pieces of equipment, the more limited the range of membrane functions. Thus, the number of different materials that can be exchanged across the cell surface, the variety of cell components that can be manufactured, the range of stimuli to which the cell can respond, and so on, are all limited because prokaryotes are as small as they can be. In other words, being small restricts the adaptability[3] of a prokaryote to changing conditions. It cramps the cell's lifestyle.

---

[3] "Adaptability" is an ambiguous word. It is used in different senses in different biological contexts, particularly when evolution is being discussed. Here we use it to mean the ability to survive changes in environmental conditions.

Prokaryotes have evolved remarkable ways of circumventing these limitations. Some bacteria change when the going gets tough into a very durable quiescent form, an *endospore* - a sort of suspended animation. They come back to life when conditions improve. Some can swim away from danger to a more comfortable environment. Some have genes that switch on

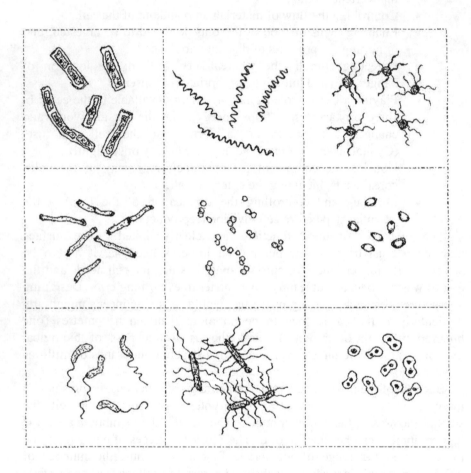

Fig. 2-2: Variety of shapes of cells belonging to just one group of bacteria (cocci).

and off in response to changing conditions, increasing adaptability. Many bacteria can exchange pieces of DNA with other kinds of bacteria, passing genes to quite different organisms. This is how antibiotic resistance has spread, creating a significant worry for the medical profession. Also, bacteria work in teams. In their natural habitats, different types of prokaryotes can assemble into mixed groups and pool their resources and capabilities for the advantage of all. These groups often have beautiful geometries.

To witness such devices is to realise that our wonder at Nature need not - indeed, should not – be restricted to the everyday macroscopic world of oaks, primroses and beetles. Nature under the microscope is wonderful and enchanting too. This is a case where the ability to explain what we observe is a major ingredient of our enchantment. Knowledge does not merely enhance pleasure; sometimes, as when we contemplate these startling assemblies of bacteria, it is essential for it.

Nevertheless, despite the remarkable devices by which bacteria cope with difficult circumstances, the limitation imposed by the small cell membrane area is intractable. Being small enables cells to reproduce at maximum speed. But it imposes severe restrictions on their individual ability to adapt.

Chapter 3

BIGGER CELLS
Eukaryotic cells and their contents

We promised not to consider evolutionary theory until later in the book, but we have already invoked it several times, mentioning biological "drive", transmission and survival of genes, adaptation, and so on. This shows how difficult it is to survey any part of modern biology without referring at least implicitly to biology's central theory. During the next few chapters, these implicit references will continue. However, we shall defer explicit discussion of the theory until chapter 11.

Our attention at present is on cell structure. So far, this seems to have brought us no closer to understanding what "life" is or why the cell should be considered the "fundamental unit of life". We ask for your patience. The question we are addressing is complicated and we have to approach our answer to it in stages.

The stage we reach in this chapter concerns cells that have overcome the basic limitation of prokaryotes: restricted membrane area. Eukaryotic cells do not have such restricted membrane areas, so they have elaborated membrane functions to a high degree. As a result, they have versatile manufacturing capabilities, they can respond to an impressive variety of stimuli, and so forth. This is only partly because eukaryotic cells have bigger surface areas than prokaryotes. (A thousand-fold greater cell volume means a hundred-fold greater surface area, if the cells being compared have the same geometries.) A more important point is that only three of the functions of the prokaryotic cell membrane *have* to be associated with the cell surface:-

- Preventing accidental mixing of the cell contents with the environment - keeping the outside out and the inside in.
- Controlling the flow of materials into and out of the cell.
- Detecting stimuli from outside and initiating appropriate responses.

The surface membrane of a eukaryotic cell is almost exclusively devoted to these three functions. So it is not only much bigger than its prokaryotic counterpart, it is also more specialised and more sophisticated. The other jobs done by the prokaryotic membrane are delegated in eukaryotes to membrane structures inside the cell.

The total area of these internal membranes can be vast. If they were confined to the surface then the length and width of – say – a liver cell would be about a millimetre - impossibly big for an active animal cell. Membrane internalisation has prevented eukaryotic cell size from becoming unmanageable. Each internal membrane system has its own functional specialism, and they all have impressive names[4]. We would not expect readers to remember these names but we need to introduce them here; the model we are going to construct would make little sense if we could not name the parts and have some idea of the functions of each part. When we mention the names again later in the book we shall remind you what they mean.

Intracellular digestion is carried out by small membrane-bound spheres called *lysosomes*. When a tasty morsel touches the outside of the cell, part of the surface membrane pinches off to enclose the morsel, forming an *endocytic vesicle*. The endocytic vesicle fuses with a lysosome, the morsel is digested, and the digestion products pass through the lysosomal membrane into the body of the cell, where they are used.

When the cell's own proteins and nucleic acids wear out, they too are broken down (digested) so that their components can be recycled. This is usually accomplished not by lysosomes but by structures called *proteasomes*. In vertebrates, proteasomes can also break down foreign proteins. In this case, the resulting fragments can be presented to the animal's immune system. If the foreign protein reappears, an immune response results.

---

[4] For readers who are interested in etymologies, here are the roots of these names and their literal meanings. *Chloroplast* comes from the Greek words *chloros* (= pale green) and *plastos* (= moulded). *Cytoplasm* comes from the Greek *kytos* (= vessel or cell) and *plasma* (= form or body). *Lysosome* comes from the Greek *lysis* (= dissolution) and *soma* (= body). All these are directly descriptive of appearance or function. *Mitochondrion* comes from the Greek *mitos* (= thread) and *chondros* (= granule), "thread-granules" describing the appearance that mitochondria presented to the 19th century microscopists who first observed them. *Endoplasmic reticulum* combines the Greek for "inside the form (body)" (*endo + plasma*) with the Latin for "little net" (*rete* = net; "*reticulum*" is the diminutive form). *Endocytotic vesicle* combines the Greek for "inside the vessel (cell)" (*endo + kytos*) with the Latin *vesica* (= bladder or blister). *Nucleus* comes from the Latin *nux* (= nut or kernel). The *Golgi complex* is named after the cell biologist who first described the structure in the 1890s, Camillo Golgi (1843-1926).

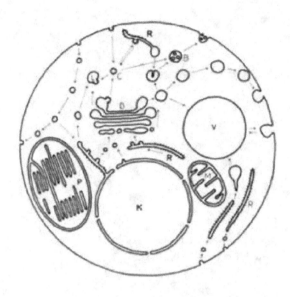

Fig. 3-1(a): schematic drawing of a generalized eukaryotic cell. K=nucleus, P=chloroplast, M=mitochondrion, R=endoplasmic reticulum (the small dots are ribosomes), C=lysosome, D=Golgi complex, B=secretory vesicle, V=vacuole. The rows of arrows to the left of the picture indicate the formation of endocytic vesicles and their fusion with lysosomes.

Energy metabolism (the provision of available energy for the cell) is largely the task of membrane-bound objects called *mitochondria*. In size, shape and many subtler features, mitochondria resemble bacteria. Far back in evolutionary history they *were* bacteria, which took up residence inside the Earth's pioneering eukaryotes and have been there ever since. We shall discuss this phenomenon in later chapters.

Trapping the energy of sunlight is accomplished by *chloroplasts*, which are found in cells in the green parts of plants and in some single-celled eukaryotes. Chloroplasts, like mitochondria, resemble certain prokaryotes (cyanobacteria); they once *were* cyanobacteria.

Many of the cell's components are manufactured by a diffuse array of membrane sheets called the *endoplasmic reticulum*. Some manufacturing and assembly tasks are undertaken or completed by a more specialised membrane system that arises from it, the *Golgi complex*.

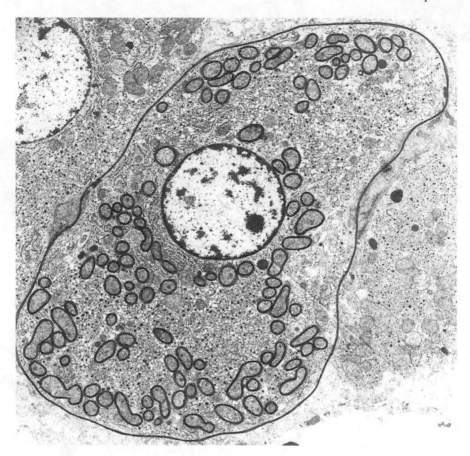

Fig. 3-1(b): electron micrograph of a sectioned animal cell. The central nucleus (the pale circle) containing a darkly-stained nucleolus, the cell membrane, and inclusions such as mitochondria (round or elongated) are clearly visible. The small black dots in the cytoplasm (the part of the cell outside the nucleus) are ribosomes.

DNA replication is initiated at the *nuclear envelope*, a double membrane separating the *nucleus* (which houses the DNA) from the rest of the cell (the *cytoplasm*). The nuclear envelope contains specialised structures known as *pore complexes* that allow RNAs and ribosomes (which are made in the nucleus) to enter the cytoplasm, where proteins are made. The pore complexes are also the sites of entry of proteins (all of which are made in the cytoplasm) into the nucleus.

Even before we construct the model of a eukaryotic cell, an obvious question arises: how are all these components kept in order? Come to that, how does an animal cell – which, unlike a prokaryote or a plant or fungal cell, has no rigid cell wall – maintain a specific shape? Since around 1970 it has been recognised that cell shape and internal order are maintained by a *cytoskeleton*. The cytoskeleton consists of fibres that occur singly, in bundles and in networks. It is not a stiff inflexible skeleton but a plastic, flexible and highly dynamic one. There are different types of fibres in the cytoskeleton. These have special names of their own, but we need not introduce them here. The cytoskeleton has other functions apart from maintaining cell shape and keeping the contents in order. For example, cell movement depends on it. So do many internal transport processes. It provides the cell with "muscle power" as well as skeletal support.

A model for a eukaryotic cell

The model we describe here is broadly appropriate for any cell from a beetle, a swallow or a human, or for a single-celled organism such as an *Amoeba*. In reality, though, animal cells vary enormously. Your body, for example, contains between ten and a hundred million million cells (10,000,000,000,000-100,000,000,000,000) and they are far from identical in size, shape, appearance or function. There are about 200 major types of human cells - nerve cells, muscle cells, liver cells, kidney cells, bone cells, gland cells and so on - and there are variations within each type. However, we are not concerned with details here, only generalities: the packing and relative sizes of the contents. So we shall overlook the differences among cell types for the time being. After we have dealt with the animal cell model we shall briefly describe a comparable model for a plant cell.

The scale we shall use is two and a half times smaller than the one we used for the prokaryote. One centimetre of model will now represent about 1 µm of cell. On this scale, the prokaryote shrinks from matchbox size to a cocktail sausage or a chocolate. The eukaryotic cell is represented by a small cardboard carton, ideally with edges of 15-20 cm (6-8 inches). This box needs to be fairly robust. As in the prokaryote model, the sharp edges and corners of the box are unrealistic. However, the most serious departure from reality is that instead of cardboard the surface ought to be very thin polythene, about a tenth of a millimetre thick, representing the cell membrane. Remember that unlike a prokaryote, an animal cell has no cell wall. Keep this in mind while you pack the box. It makes for one more surprise.

The first thing we put into the prokaryote matchbox was the DNA; so the first thing we will put into the eukaryotic cell box is the nucleus. A grapefruit or a large orange is about the right size; a roughly spherical object about 10 cm (4 inches) in diameter. In principle, we should fill the orange or grapefruit with thread to represent the DNA, but the amount of thread needed would make this impractical. Unlike prokaryotic DNA, eukaryotic DNA is not circular. If the cell we are modelling is a normal human cell, the DNA is in 46 separate pieces called *chromosomes*. The chromosomes differ in length, but if they were disentangled, stretched out and joined end to end, the total length of DNA from a single cell would be about two *metres*. On the scale of our cardboard box model, this means you would have to force about twenty kilometres (twelve miles) of thread into the grapefruit! Some types of eukaryotic cells such as yeasts have considerably less DNA, but others, such as the cells from certain species of salamander, have even more. Try to imagine how you could package this amount of thread into the grapefruit, but we would not recommend attempting it in practice.

Of course, the eukaryotic cell contains RNA, ribosomes and proteins, too. Put into the box as much thread as you can find to represent the RNA – again, in principle, several kilometres would be appropriate; half a kilo of lentils to represent the ribosomes; and half a kilo of salt for the cell's proteins. (Lentils are slightly out of scale; mustard seeds would be better, but few households contain half-kilos of mustard seeds.) There are many grains of salt in a pound (half kilogram) bag; there are many protein molecules inside a eukaryotic cell.

Now we can introduce the internal membranes and cell components that we listed earlier in the chapter:-

- two or three handsful of kidney beans or something of equivalent size - about a centimetre long - to represent the lysosomes and endocytic vesicles (the structures involved in digestion);
- half a kilo or a kilo of chocolates or cocktail sausages for the mitochondria (energy metabolism); and
- a packet of 100-200 polythene bags for the endoplasmic reticulum and the Golgi complex (manufacturing of components).

Finally, overlooking the 5 litres of water that should be included in the model, we need to add the fibres of the cytoskeleton. Probably the easiest way to represent these is with fuse-wire. Five amp fuse-wire is slightly too thick to represent the finest of the fibres (about 0.08-0.1 mm would be the right thickness) but fifteen amp (about 0.5 mm thick) is more or less appropriate for the thickest of them. Fuse-wire misrepresents the cytoskeleton because (a) the thicker fibres ought to be hollow and (b) fuse-wire bends and kinks easily while the cytoskeletal fibres are straight and unbending. Nevertheless it gives an idea of proportion. The amount you add is quite arbitrary because cells vary widely in their cytoskeleton contents, but between 20 and 100 metres of 5 amp and another 20-100 metres of 15 amp would be appropriate, if a little impractical.

We have omitted some components such as nutrient molecules, though none of these takes up much space. Also, we have been conservative about the quantities of each ingredient. Nevertheless you will probably find it difficult to close the box lid. Animal cells are much bigger than prokaryotes (remember that mitochondria, represented by chocolates or cocktail sausages, are the size of bacteria), but they are just as tightly packed.

Recall that the outside of the box, representing the cell membrane, ought to be tenth-of-a-millimetre thick polythene or cellophane instead of heavy cardboard. How does the cell stay intact rather than tearing open and spilling its contents? Two answers are usually given. First, parts of the cytoskeleton are fastened to the inside of the cell membrane; these are attached to longer cytoskeletal fibres anchored to more robust objects inside the cell. Second, the pressures on the inside and the outside of the cell membrane balance each other precisely. These are reasonable explanations; nevertheless, the stability and durability of an animal cell are remarkable.

We promised a comparable model of a plant cell. For this, use a bigger box (30-40 cm edges), put into it the same contents as you put into the animal cell, and add a fully-inflated party balloon. Ideally, the balloon should sit in the middle of the box. It represents a fluid-filled *vacuole*, or space, which occupies the central region of a typical plant cell. If the cell is in the green part of the plant there ought to be chloroplasts as well - easily simulated with gherkins. Chloroplasts are about the same sizes as mitochondria. For a plant cell, the thick cardboard of the box adequately represents the cell wall (which, like cardboard, it is normally made of cellulose); and for many plant cells the rigid geometry of the box is also fairly realistic. However, there still ought to be a tenth-of-a-millimetre thick polythene layer around the inside of the cardboard to represent the cell membrane. Once again the model is general; plant cells, like animal cells, come in various shapes and sizes; but the relative proportions of contents are about right, and so is the packing.

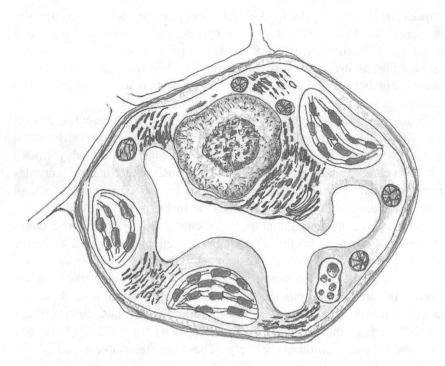

**Fig. 3-2: drawing of an electron micrograph of a sectioned plant cell.
Note the regular geometry and the thick cell wall. The irregular white
area is the central vacuole. The circular nucleus has been displaced to
the upper left part of the cell. The other large inclusions are
chloroplasts.**

Does bigger mean better?

Eukaryotic cells vary a great deal in size. Yeast cells are smaller than our
model implies; amphibian eggs are very much bigger, large enough to be
seen with the naked eye. Nevertheless the overall message is clear: even the
smallest eukaryotic cell is much bigger than the biggest prokaryote. Living
cells fall into two distinct classes, differing in size range and in contents.
There are no intermediate sizes and forms. Why not?

Eukaryotes solve the problem of the prokaryotic cell membrane (too
many jobs to accomplish in too small an area) by delegating most of these
jobs to internal structures. These internal structures can have large areas,
providing more space for more machinery. For example, the range of
materials manufactured in the cell can be enormously increased by
increasing the amount of endoplasmic reticulum. The surface membrane can
house more stimulus-detecting equipment, so the cell can respond to wider

ranges of stimuli. An increased amount of machinery means more types of proteins. Every single piece of equipment added anywhere in the cell contains at least one new protein, and there are many new pieces of equipment. This means more genes; every new protein has to be encoded in a new gene. So membrane internalisation, the eukaryote's way of elaborating and extending membrane functions, has a logical consequence: the cell must have many more genes than a prokaryote - in other words, a lot more DNA. We saw this effect in the box model: a greatly expanded total membrane area, many more proteins, a vastly increased genome, and a much bigger cell.

However, this argument does not explain why the prokaryote-eukaryote divide is so sharp. Why *are* there no intermediary forms? The only convincing answer lies in the way eukaryotes evolved, an issue that we shall defer until chapter 12.

As we saw in chapter 2, the advantage of the prokaryote's minimalist lifestyle, keeping the amount of DNA as small as possible, is fast reproduction. The less DNA there is the more quickly it can be copied (duplicated), so the shorter the time between successive cell divisions. The eukaryote pays for its greater size and sophistication by slower cell division. Cells in your body that are capable of dividing take 12-24 hours between one division and the next, rather than the 20-30 minutes typical of prokaryotes. There is so much DNA to duplicate that the job cannot be done faster.

However, by no means every cell in your body *can* divide. Most of them are specialised, concentrating on just one or two of their myriad possible functions. In specialising, they have lost their capacity to divide. Eukaryotic cell division is a very complicated process requiring dedicated machinery, and retaining this machinery is incompatible with extreme specialisation. Therefore, your nerve cells, muscle cells, secretory gland cells and all the rest of the two hundred different types do not normally divide[5]. When they wear out, they can be replaced by the differentiation (specialisation) of a stock of relatively unspecialised, dividing cells known as "stem cells"; but not always or for ever. Some day, too many of the specialist cells will wear out and can no longer be replaced, and the result is death. Whether or not death is a drawback of our multicellular eukaryotic way of life is a moot philosophical point. Individually it strikes us as highly undesirable.

---

[5] They can be made to divide outside the body by culturing them under special conditions, but with difficulty. Some cells (e.g. in your skin) can be provoked into dividing faster by injury - they produce more cells to seal the hole. In cancers, a specialised cell loses some or most of its specialisation and starts dividing again, sometimes without limit; if it becomes mobile as well, secondary cancers might develop elsewhere in the body.

Biologically speaking it is probably a good thing as well as an inevitable one.

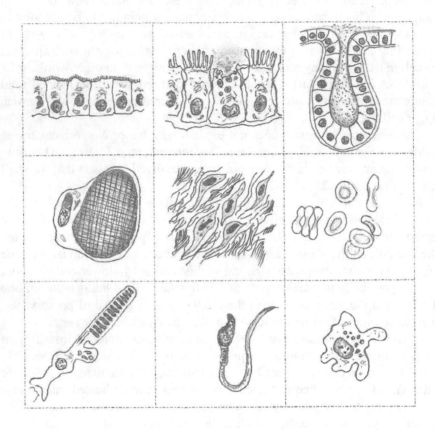

Fig. 3-3: Top row: epithelial cells: cuboidal epithelium with small microvilli; ciliated epithelium with a goblet cell (which secretes mucus) in the middle; epithelium forming a crypt in the gut lining. **Middle row:** connective tissue cells: an adipocyte from fat tissue - the dark material is a large vacuole of fat (lipid); fibroblasts among collagen fibres that they have extruded; red blood cells (erythrocytes). **Bottom row:** highly specialised cells: a retinal rod sensitively detecting photons; a sperm cell; and a phagocytic cell from the liver or blood (macrophage).

Extreme cell specialisation (differentiation) has made it possible for very sophisticated multicellular organisms such as humans and primroses to evolve. Every cell in a multicellular organism has a specific job, and the cells are arranged so that all the tasks necessary for life are reliably accomplished somewhere in the body. Not all eukaryotes develop in this way; many of them are successful single-celled organisms, with all the necessary structures and functions contained within one cell membrane. But the Earth is home to an enormous variety of multicellular plants and animals, so cell specialisation has proved an evolutionary success.

The mature form of a multicellular eukaryote is encoded in its genome as a sponge cake is encoded in its recipe. Developing from a single fertilised egg cell, the correct numbers of cells with the right specialisms are produced in all the right places; and when the numbers are right, most of the cells stop dividing. This astonishing process of embryo development is a major focus of research, and the mechanisms involved are slowly becoming clearer. We shall have more to say about this process in chapters 8 and 9.

The eukaryotic (and particularly the multicellular) way of life is more elaborate and sophisticated than the prokaryotic, but it is hard to claim that it is either better or worse. It is simply different. "Does bigger mean better?" is not, biologically speaking, a sensible question. Even working as heterogeneous teams or assemblies of organisms, bacteria cannot achieve anything like the sophistication, or the division of labour among cells, that a beetle or a fern can. But it is worth repeating that bacteria have been on the Earth a great deal longer than beetles, ferns or any of the rest of us; they outnumber us; and we depend on them.

Nevertheless we shall focus on eukaryotic cells for the next few chapters. Some of what we shall say, particularly in chapter 4, applies equally well to prokaryotes; but as we progress towards our characterisation of "life", an increasing focus on eukaryotes will become apparent.

Chapter 4

HIVES OF INDUSTRY
A survey of intermediary metabolism

The cardboard box models described in chapters 2 and 3 revealed the relative sizes of cell components. Constructing the models showed how tightly packed cells are. Also, it enabled us to talk about the functions of membranes and the relationships among DNA, RNA, ribosomes and proteins. However, the lentils and gherkins and so on that we put into the boxes were inert. The components of real cells, in contrast, are very active and dynamic. In this chapter we shall start to survey this dynamism.

Everything inside a living cell is continually moving and changing, forming and breaking down. At any instant, many or most of the cell's elementary machines - the protein molecules - are busily engaged in specialised individual activities. The proteins themselves are continually being produced and destroyed ("turned over"). At any instant, each mitochondrion, lysosome and segment of endoplasmic reticulum, every little region in the nucleus and in the cytoplasm, is buzzing with activity, each of its numerous proteins pursuing its appointed task. To describe the cell as a hive of industry would be to understate reality. The cell is a hive of hives of industry.

What sort of industry? Broadly, three kinds of processes are going on:-
- structures are being assembled and disassembled;
- chains of chemical reactions are being carried out; and
- the cell's ingredients, from water and small nutrient molecules to structures as large as mitochondria, are being moved from place to place.

In this chapter we shall focus on the second of these, the chains of chemical reactions, and we shall comment on the assembly and disassembly of structures. In chapter 5 we shall focus on the third class of processes, the cell's internal transport mechanisms.

Cellular metabolism and the complexity of life

The word *metabolism* was invented by a pioneering cell biologist, Theodor Schwann, in the mid-19th century. Like many words in biology it has Greek roots. Loosely translated it means "transformations of a heap". A cell is a huge cocktail – a heap - of different chemical substances and most of these are continually being transformed into one another.

The chemical reactions taking place in a cell can be grouped in sequences known as *pathways*. If one reaction converts substance A to substance B, another converts B to C, a third C to D and so on to Y and Z, then we speak of a "pathway" converting A to Z. Each pathway serves one of two main purposes. First, it might liberate energy from the starting material (A) and make it available to the cell. In this case Z must be a smaller and simpler molecule than A and will probably be a waste product; energy is liberated when the other molecules in the pathway are broken down to Z. Second, the pathway might produce a chemical substance that the cell needs for communicating information, building a membrane, replicating the cell, or some other purpose. In this case, Z will be larger and more elaborate than A and the A-to-Z pathway will consume energy. The first type of pathway, the breakdown and energy-liberating sort, is described as *catabolic* ("down metabolism"). The second, the synthetic sort, is described as *anabolic* ("up metabolism").

These ideas are simple but they are remote from everyday experience, so let us consider something more familiar. The food you eat is digested in your intestine. Then the products of digestion enter your blood stream and are taken to the various cells in your body[4]. Inside these cells they are used for producing energy (catabolism) or for manufacturing cell constituents (anabolism). Any nutrients that are surplus to the body's immediate requirements for catabolism and anabolism are put into storage for later use. Our bodies store carbohydrate (glycogen, which is similar to starch, is stored in liver, muscle and certain other cells), and they store fat.

It is easy to see why a growing child, many of whose cells are still actively dividing, needs more food per kilogram of body weight than an adult: she needs it to fuel the considerable energy demands of DNA replication and cell division, as well as to manufacture cell constituents. A cancer victim, who also has actively dividing cells, needs extra fuel for the same reasons but is often too ill to have much appetite. This is why cancer

---

[6] This might seem confusing because in chapter 3 we said that lysosomes digest things inside the (eukaryotic) cell. However, our lysosomes do not digest the food we eat. Our food is fully digested in the intestine, broken down into small nutrient molecules before it gets anywhere near our cells (except the cells lining the intestine). What our lysosomes digest is bacteria, debris from dead cells, and so on that are taken up in endocytotic vesicles. (Single-celled eukaryotes are different in this regard. If you were an amoeba, taking in morsels and digesting them with lysosomes *would* constitute "eating".)

patients "waste away"; body reserves and healthy tissue, particularly muscle, are broken down to provide nutrients for the cancer cells to keep dividing. Also, it is easy to see that because muscular work burns up energy, physical labour makes you need to eat more. On the other hand if you eat a lot and take little exercise you will grow fat – the excess nutrients will go into storage.

This common knowledge applies to individual cells as well as the whole body. A busy cell, one that is dividing or differentiating or manufacturing hormones for export, needs a lot of energy; so its catabolic pathways are very active. It burns fuel rapidly. A cell that manufactures a hormone must build and maintain the requisite manufacturing and secreting equipment, so the relevant anabolic pathways are active as well. A cell that takes in more nutrients than it needs for its immediate catabolic and anabolic activities converts the excess nutrient into food reserves.

Let us pause to reflect on the sizes of nutrient molecules – the products of food digestion such as glucose - and other molecules involved in metabolism. Recall the "grain of salt" image of a protein molecule in a cell (chapter 3). Proteins are *big* molecules. If a glucose molecule were magnified to the size of your body, then on the same scale a *small* protein would be the size of a double-decker bus or a terraced house. On this scale, a prokaryote would be the size of a large city and a eukaryotic cell would be a county. In terms of scale, therefore, a glucose molecule in the middle of a liver cell is like you in the middle of the county of Yorkshire, U. K. But our food produces huge numbers of these tiny molecules. If you eat a meal containing a quarter pound (about 100 grams) of starch - rice, potatoes or bread - and the starch is fully digested to glucose, then the number of glucose molecules produced is around 400,000,000,000,000,000,000,000. (This is more conveniently written 4×10^{23}, which means a 4 followed by 23 zeroes.) Even though the number of cells in your body is vast - you might remember the figure ten million million (10^{13}) - this means there are thousands of millions of glucose molecules per cell. And glucose is only one of many nutrients produced by digestion. It takes the body just one or two hours to process all this material, so an active cell metabolises *millions* of molecules of glucose (and other nutrients) every second.

These vertiginous sizes, numbers and speeds point to an ineluctable feature of the living state: complexity[7]. A glucose molecule, broken down

---

[7] "Complexity" is another ambiguous word. In its everyday sense it means the opposite of "simplicity". In modern mathematics it describes a system that borders on the "chaotic" (so sensitive to conditions that it is unpredictable) yet behaves in an ordered and stable way. Both meanings of the word apply in the present context, but for clarity, assume the everyday meaning: the living state is not simple!

by a well-defined catabolic pathway, makes its energy available to the cell. Every reaction step in this pathway (there are about two dozen steps) depends on an *enzyme*, a protein or a small group of proteins of which the sole purpose is to catalyse that reaction. Every reaction step in *every* pathway of metabolism depends on a specific, dedicated enzyme. Each enzyme in every pathway - of which there are a great many - is the product of one or more genes. All the enzymes together constitute only a fraction of the proteins that a cell makes; not all proteins are enzymes - some have quite different jobs. Your genome, the totality of your genes, potentially codes for around thirty thousand different proteins, *some* of which are the enzymes that enable all your metabolic pathways to work.

   This huge number of genes, and the corresponding number of proteins, constitutes part of the "mass of information" that is too unwieldy to make a comprehensible distinction between life and non-life (chapter 1). However, the *fact* of this complexity - the vast array of genes, the enormous variety of proteins, the bewildering network of metabolic pathways - seems to be important in itself. Part of the distinction between living and non-living might be (most biologists would say "is") that a living organism is extraordinarily complex - yet it all hangs, and works, together as an integrated, coherent whole.

Metabolic pathways
Metabolism itself is the current focus of our attention, so let us consider an aspect of metabolism that the mind can grasp: the twenty-four-step pathway of glucose catabolism. The final product of this pathway, the molecule that remains after every last drop of available energy has been squeezed out of the glucose, is carbon dioxide. This is a waste product. We dispose of it through our lungs. A glucose molecule is made up of three sorts of atoms: carbon, hydrogen and oxygen. Carbon dioxide (as the name suggests) is made up of just two sorts of atoms, carbon and oxygen. Hence, during glucose catabolism, the hydrogen atoms have been removed. What happens to them? The mitochondria (remember these are the bacteria-sized structures involved in energy metabolism – the chocolates or cocktail sausages in the box model of chapter 3) turn them into water, which is another waste product. A water molecule, H_2O, consists of hydrogen and oxygen atoms. The source of the oxygen is well known: we breathe it in. When the mitochondria combine the hydrogens stolen from the glucose molecule with the oxygen breathed in, they trap the energy released in the form of a molecule known as ATP[8].

---

[8] Details of chemistry are not crucial for this book, but just for the record, ATP is short for adenosine 5'-triphosphate: a molecule consisting of adenosine (the base adenine, which is one of the four bases in DNA, attached to a type of sugar called ribose) and a string of

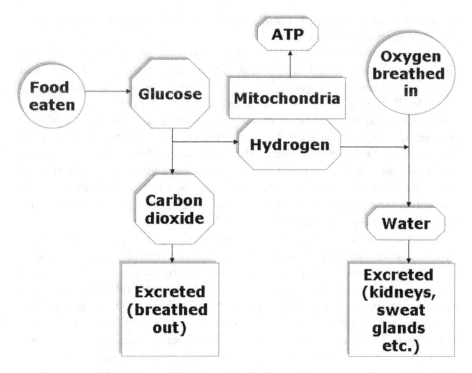

Fig. 4-1: glucose catabolism and the formation of ATP.

We have gone into some detail here, partly because this pathway of glucose catabolism is extremely widespread among living organisms, and partly because all catabolic pathways follow a similar general pattern irrespective of the initial nutrient. A molecule containing hydrogen atoms is converted to a waste material containing few or no hydrogens; the stolen hydrogens are dumped, usually (though not always) on to oxygen to form water; and ATP is made in the process. ATP provides the energy that the cell - and the whole organism - needs for synthesising and transporting materials, moving itself or part of itself, assembling structures, replicating, generating heat, sending nerve impulses, doing muscular work, and (if you

---

three phosphate groups. ATP usually makes its energy available by losing its third phosphate; the energy liberated when this happens drives a great number of biological processes including chemical syntheses. When ATP loses the phosphate it turns into ADP (adenosine 5'-diphosphate). During catabolism in the mitochondria, this process is reversed: phosphates are attached to ADP molecules so that ATP is regenerated. Readers who are interested in surprising numbers might wish to reflect on the following. A healthy human body contains about 25 grams of ATP at any instant. However, the total daily amount of ATP synthesised from ADP (and broken down again) is around 10-12 kilograms. The *turnover* of ATP is very rapid.

are a glow-worm or a firefly) producing light. It is the general all-purpose
fuel for the activities of life.

This account of catabolism leads directly to three of the seven traditional
"defining characteristics" of living organisms: eating, respiring and
excreting. These three processes are intimately linked, as the glucose
example illustrates. What we eat generates glucose; breathing supplies us
with oxygen; excretion disposes of carbon dioxide and water. We introduced
another of the traditional properties – reproduction - in chapter 2. The seven
traditional properties do not define the living state adequately, but they are
not irrelevant. So the fact that four of them have emerged so effortlessly
from our discussion suggests that we may be on the right track.   The
remaining three members of the set (movement, response to stimuli and
growth) will emerge during the next few chapters.

Every catabolic pathway involves many steps, though not necessarily
twenty-four of them; in other words, between the initial nutrient molecule
and the final waste product, there are many *intermediates*.   Some of these
intermediates are versatile molecules involved in several different pathways,
including anabolic ones. All the products of carbohydrate, fat and protein
digestion are converted inside the cell to a pool of inter-convertible
intermediates.   The intermediates can be either broken down to waste
products such as carbon dioxide and water, releasing energy for ATP
production, or used to synthesise cell constituents, consuming ATP in the
process. Fig. 4-2 shows the connections:-

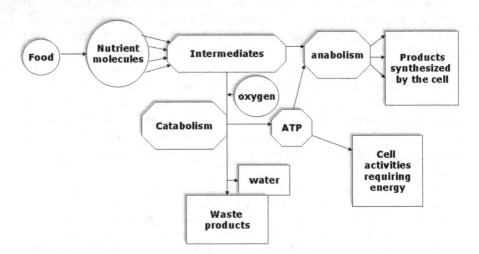

**Fig. 4-2: an overview of intermediary metabolism, showing how
catabolism and anabolism are interconnected.**

This diagram does not explicitly mention the storage compounds made from surplus nutrients. However, storage compounds are examples of "products synthesised by the cell". As we mentioned earlier, glucose can be converted to glycogen (animal starch), an important reserve fuel in many types of animal cells. In animals such as humans, excess carbohydrate can also be converted via certain intermediates to fat, which is then stored; some of us are aware of this process from depressing personal experience.

Let us recapitulate. Metabolism forges intimate links between eating, respiring and excreting. The molecules of metabolism are minute compared to protein molecules. They take part in a vastly complex array of chemical processes; complexity seems to be integral to the living state. A *metabolic pathway* is a sequence of chemical reactions, each dependent on its own specific enzyme (an enzyme is usually a protein or group of proteins). Metabolic pathways can be *catabolic* (breaking nutrient molecules down to waste products and concomitantly producing ATP, which provides energy for a wide range of life processes) or *anabolic* (manufacturing new materials from intermediates derived from nutrients). Pathways are not isolated; all metabolic pathways taking place in the same cell are interconnected.

We promise that there will not be so many new ideas in the remainder of this chapter!

Metabolism and cell structure

There is a reciprocal dependence between the components of the cell that we discussed in chapter 3, and the pathways of metabolism that we have introduced in the present chapter. Metabolic pathways depend on cell structure and organisation, and cell structure and organisation depend on metabolism.

In mitochondria, the enzymes necessary for certain stages of catabolism are lined up on the membrane like little workstations along a conveyor belt. The starting material, a metabolic intermediate, is chemically converted by the first enzyme "workstation". The product of this conversion hops directly on to the second enzyme, which is held in an immediately adjacent position; and so on for enzyme after enzyme. This arrangement ensures that the pathway is rapid and efficient. (Changing trains at a succession of railway stations would be like this if the trains ran on schedule and at times - and to places - that suited your needs.) In the endoplasmic reticulum (the membrane system that is concerned with manufacturing processes), similar enzyme arrays ensure the speed and efficiency of anabolic pathways. The molecule being transformed is passed from enzyme to enzyme in the correct sequence and with the minimum of fuss. If the enzymes were not appropriately aligned on the membranes, then speed and efficiency would be lost – perhaps fatally.

The enzymes of some metabolic pathways are not bound to membranes but apparently "free" in the cytoplasm. "Free" is not a literal description, however; they are often linked together in loose assemblies called *metabolons*, in some cases possibly linked to the cytoskeleton. Individual metabolons are not as durable as membranes but they are another way of organising arrays of enzymes. As in membrane-associated pathways, the molecule being processed is transferred from one enzyme to the next with maximum efficiency. Generally, therefore, the efficiency of cellular metabolism depends on cell structure and organisation.

On the other hand, how are structures such as mitochondrial membranes and endoplasmic reticulum and cytoskeleton and metabolons built and maintained? Their component molecules are – obviously - manufactured by the cell. Their proteins are made by ribosomes, using the instructions on the messenger RNA "photocopy" of the gene. All their other components (lipids and complicated carbohydrates) are made by anabolic pathways. To make or replace any of these components, an anabolic pathway is necessary. All anabolic pathways need ATP, and so does protein synthesis at the ribosomes. So catabolic pathways are necessary as well; there is no other source of ATP. In short: to assemble a cellular structure from its components, or even to disassemble it in a controlled way, metabolites and metabolic pathways are necessary. Cell structure and organisation depend on metabolism.

There is a subtler point: the environment inside a cell can be surprisingly destructive. Chemical derivatives of oxygen can irreversibly alter important molecules, rendering them useless. Therefore, the cell has to protect its molecules against the continual threat of chemical damage. The protective devices that it uses are products of metabolism. Therefore, metabolism is needed to protect and maintain the cell's structures as well as to make, replace, assemble and disassemble them.

Cell structure and organisation are necessary for metabolism and metabolism is necessary for cell structure and organisation. In our view, this absolute and intimate interdependence is part of the essence of "livingness".

Back to Nature
Readers who are interested in wildlife rather than cells or molecules, things that can be seen with the naked eye rather than with sophisticated laboratory equipment, might find this account dissatisfying. However, an understanding of metabolism can help us to appreciate the rural idyll we evoked in chapter 1. Most people know that plants use sunlight to manufacture food (glucose and starch) from carbon dioxide. The process is called *photosynthesis*. The cell structures involved in photosynthesis are the

chloroplasts (the gherkins in the model in chapter 3) Like all metabolic processes, photosynthesis is complicated, but simplified scheme of it looks like this:-

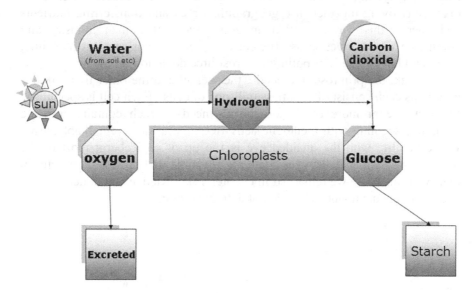

Fig. 4-3: photosynthesis. Note the comparison with Fig. 4-1.

This is almost a double-mirror-image of the scheme of glucose catabolism shown earlier in this chapter. Some of the sunlight energy falling on the green parts of the plant is trapped in the starch. When we eat the plant and digest the starch, our cells catabolise the resulting glucose, making this energy available as ATP, which fuels our life processes. Thus, our body's energy comes indirectly from sunlight. Moreover, the diagram shows that the plant's raw materials, water and carbon dioxide, are the excreted waste products of our catabolism. In return, the plant excretes a waste product, oxygen, that is essential for us. This is typical of the natural world: one organism's waste is another one's food. Ultimately, little is wasted.

The oak tree uses the sun's energy to turn carbon dioxide and water into food. Aphids feed on the sap in the oak's trunk, exploiting this food. The beetle crawling down the trunk eats the aphids. It digests them and metabolises the digestion products to fuel its activities, which include aphid-hunting. If the weasel eats the beetle, it will metabolise the products of digested beetle. When the weasel excretes waste, or dies, various fungi and bacteria in the soil fuel digest weasel excrement or dead weasel. The waste products of these bacteria and fungi include carbon dioxide and water, along with simple nitrogen compounds that the oak can absorb through its roots.

The oak uses these simple compounds to make food for itself - and for aphids.   Thus, energy and materials are endlessly recycled among the organisms we see around us (or cannot see because they are microscopic), and the recycling process is solar powered.   A collection of different organisms living together in a geographical area and transferring nutrients and energy this way is called an *ecosystem*.   If human industry and commerce were a fraction as efficient as a natural ecosystem at recycling, we would have no serious pollution or resource depletion problems.

The oak, the primrose, the beetle, the weasel and the other living things around us each consist of countless millions of cells.   Each cell is specialised to meet one or more of the organism's needs.   Each contains intricate structures seething with metabolic activities.   Even more remarkably, what we see of life with the unaided eye is the tip of an iceberg; most of the organisms essential to the ecosystem are invisible, but no less wonderful in their workings.   As we remarked in chapter 1, knowledge and understanding add to our enchantment.   They do not detract from it.

Chapter 5

DELIGHTS OF TRANSPORT
How the cell's contents are moved around

In chapter 4 we introduced *metabolism*, the chains of chemical reactions that take place inside living cells. We emphasised the close reciprocal dependence between metabolism and cell structure. Now we turn to another major aspect of cell activity: transport. How are a cell's ingredients imported, exported and moved from place to place, and how are its internal structures kept in position? And how do cells move?

A variety of transport mechanisms
The contents of a eukaryotic cell range from the tiny molecules of metabolism to large internal membrane structures. A protein molecule is around ten times longer than a metabolite molecule. A molecule of RNA is hundreds of times longer than a protein. A mitochondrion is three or four times longer still, and a great deal fatter than an RNA molecule. Cargoes of such widely different sizes are unlikely to be moved efficiently by a single mode of transport. Therefore, the cell has a variety of transport mechanisms.

The main ingredient of a cell is water. An average-sized human cell contains up to 1,000,000,000,000,000 (i.e. 10^{15}) water molecules. Picture a small lake fed by fast-flowing streams and drained by a large river; the level of water in the lake remains practically constant, and underwater currents flow, but no individual water molecule stays in the lake for long. The flux of water in a cell behaves similarly; water enters and leaves all the time and currents flow, but the amount inside the cell scarcely changes. But in a cell, unlike a lake, the "feeder streams" and "outflows" are dispersed all over the surface. Water flows continually in both directions across almost the whole of the cell membrane. Similarly, it enters and leaves each of the membranous inclusions - nucleus, mitochondria, lysosomes and so on.

Within the cytoplasm (the part of the cell outside the nucleus), water flows first one way and then another, or round in circles. These movements are generated in various ways: by activities of the cytoskeleton, transport of metabolites across membranes, and metabolic production and utilisation of water. (Remember that water is a product of catabolism[9]).

In chapter 4, we saw that metabolic pathways are efficient because the enzymes are fixed in ordered sequences on membranes and metabolons. Their efficiency is further enhanced by the flow of water inside the cell, which continuously feeds raw materials to the enzymes and removes the products. (Most nutrients and metabolic intermediates are soluble in water.) The principle is familiar to chemical engineers: making the reactants flow over a catalytic bed ensures that the product is made quickly and in high yield. Living cells discovered how to put this principle into practice more than a thousand million years before chemical engineers existed.

The cytoplasm is not a simple liquid, even if we discount the cytoskeleton and all the internal membrane structures. If - to revert to the box model of chapter 3 - the half-kilo of salt representing the cell's proteins had been mixed with the other solid ingredients and the appropriate volume of water (5 litres), a runny paste would have resulted. Salt is not protein but the effect is similar. Proteins are sticky molecules; they adsorb water and they adhere to each other. So the cytoplasm, minus membranes and cytoskeleton, can be pictured as a runny paste. Examined under a high-voltage electron microscope it is a loose network of thin strands, mainly water-saturated proteins. However, cytoplasm is not a stable gel like a table jelly. The strands are continually breaking and reforming; those in a table jelly are much less labile.

Water, and the small metabolite molecules dissolved in it, flow through the cytoplasm fairly easily and quickly. Bigger molecules are a different matter. They repeatedly become entangled in the network, and however quickly the strands of the network break and reform, this slows their movement. For protein and RNA molecules, which tend to stick to the network and become part of it, the slowing is potentially dramatic. They could be almost immobilised unless a path could be cleared for them.

---

[9] A simple calculation can be made, based on the amount of oxygen your body takes up per unit time while you are at rest (sitting in a chair reading this book, for example). If we assume that nearly all the oxygen you breathe in is converted to water by catabolism in the mitochondria; your body contains ten million million cells; and an average cell contains a hundred mitochondria... then ten thousand water molecules are produced in each mitochondrion every second. If you exert yourself, then you take in oxygen more rapidly and the rate of water production increases.

Many proteins, and probably messenger RNAs, are targeted for specific destinations. A protein might be destined to remain in the cytoplasm, or to become part of a mitochondrial or lysosomal membrane, or to take up residence in the watery space inside a mitochondrion or lysosome. It might be dispatched to the cell surface or to the nucleus. It might be exported from the cell into the outside world. The cell has to ensure that the right proteins go to the right destinations. In principle, a newly-made protein could wander around the cell at random more or less indefinitely and then bind anywhere; so how is this avoided? The cell uses its proteasomes (see chapter 3) to hoover away proteins that "hang around" for more than an hour after synthesis and fail to find their destinations. This would be fatally wasteful unless there were efficient, protective cytoplasmic transport processes for large molecules.

In many instances, one part of protein molecule functions as a "travel ticket", which is recognised by a "ticket inspector" at the target site but is not valid for other destinations. This ensures that the protein stops at its ordained destination, but does not explain how it travels there. The cytoplasmic transport processes for proteins and RNAs are certainly efficient, but the mechanisms are not clear and the subject is a matter of controversy. In some cases, movements of the cytoskeleton might drag a protein along (see below). In other cases a big molecule might be passed from one strand of network to the next in a series of little jumps, perhaps with an energy-requiring (ATP-dependent) "push" to start it at source. In still other cases, big molecules might become attached to membranes and carried along by membrane flow (again, see below). And there might be other mechanisms of which we are currently ignorant.

Among some biologists, the belief persists that proteins travel in the cytoplasm by diffusion. This idea is superficially attractive because diffusion is a simple physical process, the result of random thermal motion of molecules, and requires no specially evolved apparatus. However, random molecular saccades do not provide a plausible basis for specifically directed movement over cellular distances, particularly in an environment where random motion would be seriously retarded by the cytoplasmic network.

Molecules that are insoluble in water present different problems. They cannot move by exploiting water flow in the cell. One way of transporting water-insoluble molecules is to "dissolve" them in the endoplasmic reticulum (membranes are greasy rather than watery, so they are friendly environments for water-insoluble molecules). They flow slowly along the planes of this membrane system to reach the cell surface, the Golgi complex, the lysosomes or other destination. Membranes flow rather as a slick of oil flows over water; slow movement along a plane. Just as an oil slick flows when oil is added to one side, so a membrane flows when new material is added by anabolic processes. And just as an oil-soluble substance added to the puddle will be borne across the surface in the flow of oil, so a water-insoluble molecule in a cell is borne along by the flowing membrane. Some water-insoluble molecules, however, travel by a different method. They bind to a protein that is targeted to the appropriate destination. This enables them to cross the cytoplasm under protection, much as a traveller might cross the desert by attaching himself to a camel train.

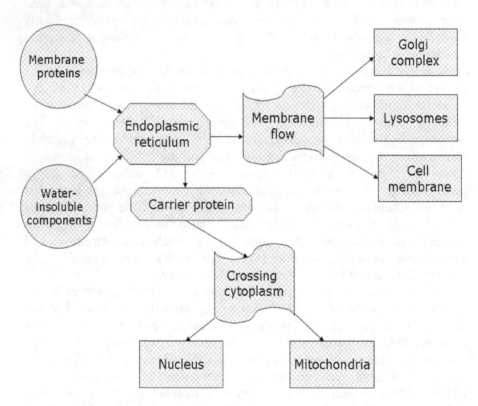

Fig. 5-1: how proteins and water-insoluble components are transported to the various compartments of a eukaryotic cell.

   Membranes therefore act as transport routes for some water-insoluble materials. But they are barriers for water-soluble ones; the first job of the surface membrane is to stop the inside of the cell mixing with the environment. To move, say, a glucose molecule into the cell from the outside, a special piece of equipment is incorporated into the cell membrane. Many protein components of membranes are just such pieces of equipment;

devices for transporting or pumping[10] particular water-soluble molecules from one side to the other.

Let us sum up so far.

- Water flow inside the cell is generated by metabolism, by membrane transport processes, by energy-dependent movements of the cytoskeleton, and perhaps by other means.
- Small water-soluble molecules move through the cytoplasm in the flowing water. They cross membranes by means of specific transport devices or pumps. Metabolite flow over enzyme (catalytic) assemblies probably increases the efficiency of metabolism.
- How large water-soluble molecules move through cytoplasm is not completely understood, but proteins are often targeted to specific destinations where they become bound.
- Water-insoluble membranes either "dissolve" in the endoplasmic reticulum and are carried by membrane flow to their destinations, or hitch lifts on proteins traversing the cytoplasm to particular targets.

Transport and the cytoskeleton

The fine structure of the cytoplasm hinders the movement of large molecules such as proteins, so bigger objects such as lysosomes and mitochondria must be more or less immobile. Indeed they are, unless the cytoskeleton lends a hand.

A cytoskeletal fibre is built like a popper-bead necklace but it is more rigid. The "beads" are special sorts of protein molecule designed to fit together to form "necklaces".

The fibre can be lengthened by adding more "beads" and shortened by removing them. The lengthening (assembly) process costs energy (ATP). If beads are added at one end and simultaneously removed at the other end then the whole fibre appears to move. (Try it with a popper-bead necklace.) If the growing end of the fibre is attached to the cell membrane, the entire cell is moved as a result: the extending fibres push out the membrane, rather as you might push out the finger of a rubber glove, and the cell contents follow. Amoebae travel by this method; so do some cells in your body. If an internal membrane structure such as a mitochondrion is caught among assembling/disassembling fibres, it will be pushed from one place to another despite the resistance of the cytoplasmic gel. Perhaps such processes also move certain types of protein and RNA molecules around inside the cell.

---

[10] In some cases the equipment for transporting a water-soluble molecule across a membrane requires energy, supplied directly or indirectly by ATP. An energy-requiring membrane transport machine is commonly called a "pump".

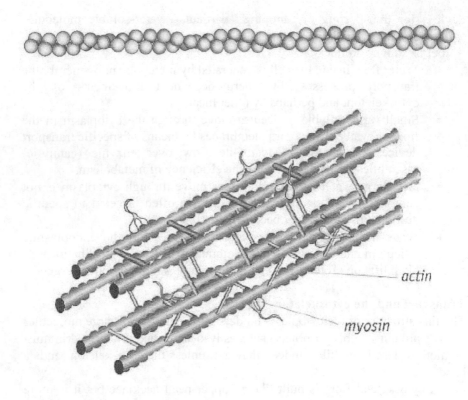

Fig. 5-2: upper drawing - one type of cytoskeletal fibre showing the arrangement of the component protein (actin) molecules; lower drawing - the arrangement of such fibrils together with the contractile protein myosin in a muscle cell.

However, the assembly and disassembly of cytoskeletal fibres do not normally move large intracellular objects such as mitochondria. A more usual method involves *motors* associated with the cytoskeleton. A motor is an ATP-fuelled molecule that runs along the fibre and is attached to the object to be transported. The system is rather like a goods train on a monorail. Remarkable distances can be travelled by this mechanism. For example, neurotransmitters (chemicals that are released from the end of a nerve cell when an electrical impulse arrives) are packaged in tiny membrane-bound vesicles. These vesicles are made in the body of the nerve

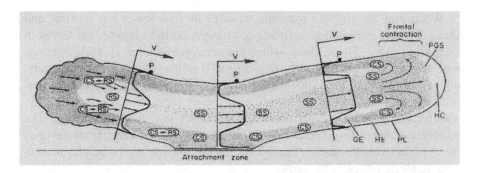

**Fig. 5-3: a cell undergoing 'amoeboid movement' caused by the co-
ordinated assembly and disassembly of cytoskeletal fibres.
SS=stabilised (non-moving) regions, CS=contracted regions, RS=relaxed
regions, HC=cap (leading edge of cell), V=direction of movement,
PL=cell membrane, and GE, HE and PGS are regions of cytoplasm in
different states of rigidity and fluidity. Drawing taken from de Robertis
and de Robertis *Cell Biology*.**

cell and transported to the end of that cell, which might be surprisingly far
away. The muscles that make your toes move are in the lower part of your
leg. The nerves that control these muscles have their cell bodies in your
spine, so the neurotransmitter vesicles have to be carried all the way down
your leg to reach the ends. (Of course they are already in place when you
wiggle your toes.) The *axons* of these nerve cells, the parts that carry the
electrical impulses, extend this entire distance. The neurotransmitter
packages are carried by a motor-driven mechanism along cytoskeletal fibres
the whole length of the axon.

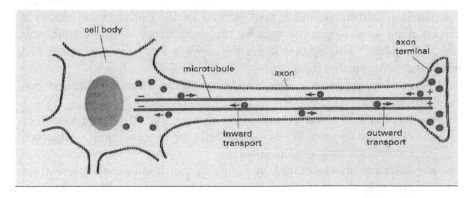

Fig. 5-4: axonal transport in a neuron (see also chapter 16).

When you wiggle your toes, the muscles in your lower leg contract and relax. Muscle contraction involves specially-adapted cytoskeletal fibres in the muscle cells. These fibres, which are arranged in parallel, slide between one another, shortening the muscle cell when they slide one way, lengthening it when they slide the other. The sliding of the fibres is another motor-driven mechanism. All animal muscles seem to work in this way: your biceps, a blowfly's flight muscles, a shark's jaw, a worm's body. The cytoskeletal fibres involved in muscle contraction are different from the ones involved in neurotransmitter transport, but the underlying principle, motors running along fibres, is the same.

Eukaryotic cell division, the basis of growth, also involves motors on cytoskeletal fibres. In this case the paired duplicate *chromosomes* are separated by motor-driven fibres so that each daughter cell receives an equal share. A chromosome is a single DNA molecule packaged with specialised proteins. Remember there are 46 DNA molecules and therefore 46 chromosomes in an ordinary human cell. They must all be duplicated before the cell divides so that each daughter cell still has the correct chromosome number - 46. The motor-driven fibres separate one set of 46 from the duplicate set; then the cell splits into two identical, viable halves.

There is another mechanism of cell movement, found in some prokaryotes and single-celled eukaryotes. Long extensions resembling whips (*flagella*) protrude from the cell, and the rhythmic beating of these generates a swimming action. This rhythmic beating is another motor-driven process. In some cases the flagella describe circular movements; bacteria invented the wheel more than a thousand million years ago! Some cells in multicellular eukaryotes have rather similar projections from their surfaces. These projections, known as *cilia*, are extensions of the cytoskeleton and their movements are once again motor-driven. Co-ordinated movements of cilia along rows or sheets of cells make the surrounding fluid move, rather than the cell itself. For instance, mucus is driven along your respiratory tract by the movements of cilia on the cells lining the tubes. This process keeps the airways free of contaminants that have become trapped in the mucus.

In summary, we can make three additions to the list of cellular transport mechanisms surveyed in the first section of this chapter:-

- Cytoskeletal fibres assemble and disassemble. In the process they cause the movements of whole cells, or – sometimes - large cell ingredients such as mitochondria.
- Cells can also be moved, or can make the fluid in contact with them move, by the rhythmic motor-driven beating of flagella or cilia.
- Motors moving along cytoskeletal fibres can be used to move or transport membranous structures, chromosomes, other fibres, and perhaps some of the cell's larger molecules.

Transport, metabolism, structure and organisation

We have now surveyed a variety of mechanisms underpinning the fifth traditional property of organisms: movement. This is almost incidental. More significantly, our discussion leads to further insights into the nature of the living state. At the end of chapter 4 we drew attention to the interdependence between metabolism and cell structure and suggested that this was the first step towards answering the question "What is the fundamental difference between living and non-living?" We can now take a further step.

Early in the chapter we noted that transport is essential for metabolism. Clearly, if enzymes fail to reach their destinations, they cannot be incorporated into the "assembly lines" responsible for metabolic pathways. This would make the pathways non-functional; metabolism will break down. Protein transport is therefore necessary for metabolism. Moreover, metabolism depends on nutrients entering the cell and on metabolites entering compartments such as mitochondria where they are processed; so transport across membranes is essential for metabolism. The flow of water in the cytoplasm ensures that metabolic processes are efficient. In these ways, *metabolism depends on transport.*

On the other hand, many transport processes require energy. They would cease if the cell did not supply ATP. Metabolism (specifically, catabolism) is necessary for the supply of ATP. Moreover, the equipment required to transport materials across membranes has to be synthesised in the cell, so it depends on anabolic pathways. In short, *transport depends on metabolism.* We suggest that the interdependence between metabolism and transport is as profound, and as integral to understanding the living state, as the interdependence between metabolism and cell structure.

There is a similar interdependence between transport and cell structure. Membrane flow carries water-insoluble molecules towards their destinations. The cytoskeleton transports large cell components. So transport often depends on cell structures. On the other hand, the cell's orderly and efficient transport mechanisms are necessary to deliver building materials to the structures for which they are destined. They are also needed to transport these structures to sites where they are needed. Thus, cell structure and organisation depend on transport; transport depends on cell structure and organisation.

In short: *metabolism, transport and cell structure and organisation all depend on one another.* These reciprocal relationships are characteristic of and, we believe, fundamental to the living state. It is easier to accept this in relation to eukaryotic cells than prokaryotes; the small size of prokaryotes makes transport processes hard to study experimentally. Nevertheless the same reciprocity seems to apply in prokaryotes, though the range of actual processes involved is narrower than in eukaryotes and many mechanisms are simpler. (Prokaryotes have much less elaborate cytoskeletons, for example.)

The three-way interdependence among cell structure, metabolism and transport is *part* of our characterisation of the living state, but it is not the whole of it. As yet, we have mentioned genes and gene expression only incidentally, and we have made only passing references to the responses of cells to environmental stimuli. We shall start to address these topics in chapter 7; but first, it is time to review our picture of the cell to date.

Chapter 6

AS IF STANDING STILL
Cellular homeostasis and regulatory processes

The problem of control

During the previous chapters we have started to build up a picture of life at the cell level. The picture is not yet complete but it has progressed sufficiently for the reader to see a problem: how is the cell kept in order?

Within the tiny space of the cell, many different structures are tightly packed. These structures are continually being made, repaired, broken down, recycled and moved from place to place. Thousands of different types of proteins are being synthesised all the time, moved to specific destinations, used, and finally degraded. Some of these proteins are components of the many different membrane systems; some form loose fibrils in the cytoplasm, others are components of the cytoskeleton, still others are located inside the nucleus or the mitochondria or other membrane-bound compartments. Numerous metabolic processes are taking place simultaneously in all compartments of the cell, usually at dizzying speeds, each separate individual reaction requiring its own enzyme. And everything is continually in flux, from cellular water movements and membrane flow to the motor-driven activities of the cytoskeleton. Yet the cell appears to be calm and orderly. Cells can change in form and function, they can divide, they can die; but often they seem to remain essentially unchanged for long periods of time. In view of the mob of unruly components of which they are made, the potential for uncontrollable chaos, how is this apparent constancy achieved?

Modern biology's answer to this question is probably incomplete. In so far as an answer is available, we shall not be in a position to do it justice until chapter 9. We raise the question at this stage because we need to explore the idea of "internal state" before we can develop our picture of the cell further.

Homeostasis

The problem of maintaining order and apparent constancy was addressed at the whole-body level in humans and other mammals long before it was seriously considered at the cell level. The classical physiologists of the 19th and early 20th centuries discovered much about the workings of the human body. Amongst their achievements was the discovery that many of the body's measurable properties remain more or less constant even when they might be expected to change. For instance, healthy individuals maintain a steady blood pressure irrespective of whether they are lying down, standing up, walking or exercising vigorously, though the heart rate is markedly different in these four situations. The constancy of the blood pressure is explained as follows. Some major arteries contain pressure sensors. These sensors send messages along nerves to part of the brain, which then alters the heart rate and the diameters of some blood vessels. The effect is to keep the pressure within narrow limits. If the sensors detect a fall in arterial blood pressure, the heart speeds up and the vessels contract. These changes counteract the fall. If the sensors detect a rise, the heart slows down and the vessels dilate to bring the pressure back to normal. In other words, blood pressure is maintained by a feedback control system analogous to the thermostatic control of room temperature.

Control of blood pressure is just one of many examples of the feedback principle in physiology. Others include control of body water content, blood glucose concentration, the levels of oxygen and carbon dioxide in the blood stream, and body temperature. In each case the relevant parameter is maintained within narrow limits. If it climbs too high, control mechanisms reduce it. If it falls too low, control mechanisms increase it. Sometimes the adjustment is brought about by nerves that carry messages to and from the brain, as in the case of blood pressure. Sometimes it is achieved by the actions of hormones, as in the case of blood glucose level. (The best-known of these hormones, insulin, decreases the blood glucose level; a complementary hormone, glucagon, increases it.) In either case the essential principle is feedback: a sensor detects change and a control system counters it. Pondering the generality of this principle when he was an old man, the great French physiologist Claude Bernard famously remarked that "the constancy of the internal environment is a precondition of life".

The "internal environment" is the environment in which the body's cells live. Unless this environment stays nearly constant the cells will die. In the late 1920s, the American physiologist Walter Cannon invented a word to denote the constancy of the internal environment and the mechanisms responsible for maintaining it: *homeostasis*. Once again, the roots of the word are Greek: "homeostasis" means "as if standing still". The study of homeostatic mechanisms, the control of parameters through feedback, has become a large part of physiology, and Cannon's new word has entered the vocabulary of science.

Physiological variables do not *actually* stand still; they just *appear* to do so. For example, every litre of your arterial blood contains about a fifth of a gram of oxygen. If it contained significantly less, or significantly more, you would be seriously unwell. If you are resting rather than exerting yourself, it will take about six minutes for the mitochondria in your cells to turn that fifth of a gram of oxygen into water (remember that mitochondria do this in the process of making ATP). But in six minutes' time, and in an hour's time, and in a year's time, a litre of your arterial blood will still contain a fifth of a gram of oxygen, provided you keep breathing. Your cells will continue to use up oxygen and you will continue to breathe it in, and the level in your blood stream will stay more or less the same. When you exert yourself, running upstairs or lifting a heavy weight, your muscle cells consume oxygen more quickly. They have to make more ATP per second when they are working harder, so their oxygen demand rises. But you breathe more rapidly and deeply to compensate, so a litre of your arterial blood still contains about a fifth of a gram of oxygen. When your body's oxygen demand falls, as it does during sleep, your breathing slows down; once again, the same blood oxygen level is maintained. This is another example of homeostasis that depends on feedback.

At the start of chapter 5 we used the analogy of a small lake fed by streams and draining into a river. The water in the lake is continually changing but the level remains constant. This could serve as a metaphor for homeostasis.

Cellular homeostasis
Every cell in the human body must preserve a constant volume and water content. Many other parameters must also be kept constant: pH, sodium and calcium contents, ATP level, and so on. Too low an ATP level would be disastrous because many ATP-dependent processes would cease, including cell volume maintenance. Too high an ATP level would be just as bad; several key metabolic reactions would be switched off and some cellular structures would start to disintegrate. The same applies to other parameters that need to be kept within tight limits, and in all these respects the cell has to fend for itself. The homeostatic mechanisms of physiology take care of the body's internal environment, the world in which the cells live. But each individual cell still has to maintain the constancy of its own interior. This applies to *all* cells, not just those in the human body; all animal and plant cells, and all single-celled organisms, prokaryotes as well as eukaryotes. Indeed, a moment's reflection will tell you that a cell has to do an enormous number of tasks to regulate its own activity – after all, it has no other

"authority" to turn to! Numerous physical and chemical features of the interior of any cell must be kept within narrow bounds if the cell is to survive.

How is constancy within cells achieved? Several types of mechanism are involved. Three can be mentioned here, because they depend neither on changes in gene expression nor on stimuli from the environment. They depend entirely on the aspects of cell life that we have discussed so far in this book: structure, metabolism and transport.

One depends primarily on cell structure. Metabolites and the enzymes that process them might be located either in the same cell compartment so that they can interact, or in different compartments so that they are kept apart. Every step in metabolism depends on the metabolite having access to the appropriate enzyme, so this is a fairly crude but nonetheless effective way of controlling metabolic pathways – either enabling them to function or preventing them from functioning.

The second type of mechanism is inherent in metabolic pathway design. Every enzyme in a pathway contributes to the control of the pathway's overall rate. This topic has attracted the attention of mathematical biologists and a well-established body of theory has resulted, throwing light on some otherwise puzzling experimental data. Curiously, many biochemists pay scant attention to this body of theory and maintain that only certain "key" enzymes, which are subject to feedback control (e.g. by the end product of the pathway), contribute to metabolic rate regulation. "Key" enzymes probably help to determine the balance between alternative pathways, but in general they contribute no more to the rate of a *particular* pathway than any other enzyme in that pathway.

The third type of mechanism depends on transport rates. If nutrients and metabolites are supplied more rapidly, the rate of each metabolic pathway that uses them will increase. Slower supply rates mean slower metabolism. Not much research has been done on this aspect of cellular homeostasis, so details are lacking. But it might be unwise to underestimate its significance.

This picture of "cellular homeostasis" is of course incomplete. Changes in expression of genes, and responses to stimuli from outside, can modify the behaviour of a cell quite dramatically. We shall turn to these topics in the following three chapters.

Internal state
The cells of a multicellular organism such as a human take on a wide variety of roles and appearances. Taken together, all the primroses, beetles, and millions of other species of eukaryotes alive today present a bewildering range of cell types. In addition, there are the prokaryotes. Such is the variety of cell forms and functions that no single *concrete* definition or

characterisation of "livingness" is possible. An acceptable cell-biological answer to "What is life?" has to transcend description; it must be *abstract*. Moreover, as we have seen, living cells are vastly complex, so to describe any cell fully would require an impossible amount of detail. A manageable definition or characterisation must cut through this morass of specifics to generalities; so our answer must be general as well as abstract.

We have already used a number of general terms. The phrase *cell structure and organisation* denotes all the membrane systems of the cell (mitochondria, lysosomes and so forth), the numerous fibres of the cytoskeleton, the ribosomes, chromsosomes, metabolons and other multi-molecular constituents, and their relative dispositions in space. *Metabolism* encapsulates all the anabolic and catabolic pathways in the cell. If these were written out in detail the result would resemble a street map of a large city filled with moving traffic. *Transport* covers a variety of mechanisms by which a cell's ingredients, from water and small metabolite molecules to objects as big as mitochondria, are moved from place to place. Only by describing the cell in terms of these generalities has it been possible to show that *cell structure and organisation, metabolism and transport are interdependent*. This was the conclusion of chapter 5. In the present chapter we have seen that in ways that are not yet entirely clear, *this interdependence is partly responsible for the apparent stability and constancy of the cell's overall appearance and behaviour*. Cellular homeostasis seems to be rooted (at least partly) in the three-way relationship among transport processes, metabolism and cell structure. We need a convenient term for this three-way relationship and the homeostatic control that it generates.

From now on we shall describe the picture of the cell encapsulated in Fig. 6-1 as the cell's *internal state*. "Internal state" denotes the quantities of all the cell's ingredients at a particular moment, their organisation in space, the sum total of the metabolic events taking place, the directions of all the transport processes and what they are transporting, and the ways in which these different features interlock. "Internal" indicates the overall situation within *one cell*, not what might be taking place outside it or in other cells. "State" is fairly non-committal, but connotes a definable, reasonably stable situation. When we use the expression "internal state", think of the following diagram, but bear in mind that each apex of the triangle hides a huge wealth of descriptive detail.

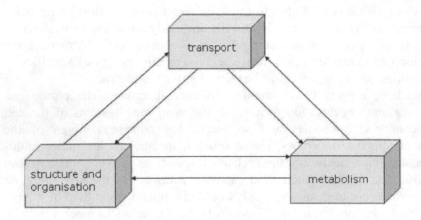

Fig. 6-1: the internal state of a cell.

"Internal state" is a summary description of a cell *at a particular moment*. In principle, it can change from one moment to the next. In practice, the internal states of most cells remain approximately constant for extended periods. However, "at a particular moment" should be regarded as part of the definition. The reason for this will become clear in the following three chapters.

The concept of internal state, which often appears to "stand still" but never does, is an important part of the answer to "What is life?" But it is not the whole answer. This is clear from the fact that we have not yet covered some of the traditional properties of organisms (*response to stimuli* and *growth*) and we have said little or nothing about reproduction or DNA or gene expression. We could not seriously tackle these aspects of biology before establishing the concept of internal state. Now that we have defined this concept, we are in a position to address them.

Before we do so, we shall look briefly beyond the cell and the single organism to the ecosystem. Does the concept of homeostasis that we have introduced in this chapter have any relevance in ecology?

"Homeostasis" in ecosystems?
In chapter 4 we suggested that the ideas of metabolism could be extrapolated from the individual cell or organism to a whole ecosystem. Energy and

materials are passed in sequence between the component organisms of the ecosystem and the inorganic parts of the environment (air and soil); there is efficient recycling. This is a well-established aspect of ecology. However, whether the idea of homeostasis can be extrapolated to ecosystems is more controversial.

The population of any organism depends on the populations of other organisms in the ecosystem, particularly those that it eats (prey) and those that eat it (predators). Populations of different species are therefore interrelated. They also depend on environmental factors such as temperature and sunlight. Because of these connections, ecosystems behave as though they had internal control mechanisms: change one population level or relevant physical parameter, and the rest of the ecosystem will respond (within limits) so as to resist the change and restore the *status quo*. However, nothing obviously analogous to the feedback control processes in physiological homeostasis can be found at the ecological level, so to use the word "homeostasis" in this context seems dubious.

Many proponents of the "Gaia Hypothesis" think otherwise. Strictly speaking, the Gaia Hypothesis merely holds that life affects the non-living environment, just as the environment affects life. Computer models such as "Daisyworld" illustrate the idea. Briefly: imagine the surface of a planet heated by a sun and populated solely by black and white daisies. Suppose black daisies grow faster at lower temperatures and white ones at higher temperatures. Black daisies absorb solar radiation and heat up the planetary surface; white ones reflect radiation so the surface cools down. Left to its own devices, this planet will settle down to a balanced population of black and white daisies maintaining a steady surface temperature. This is an interesting observation, and more complicated computer simulations have added to the interest, but it is stretching things to call such phenomena "homeostasis". We can accept "Daisyworld" as a simplified analogue of terrestrial processes without claiming that the whole biosphere is a giant ecosystem behaving homeostatically.

This topic is of general interest, not least because of the likelihood that contemporary human activity is radically changing the Earth and its biology. It cannot be dismissed out of hand. However, because our focus at present is on cells rather than ecosystems and global ecology, we shall not discuss the "Gaia Hypothesis" further until later in the book.

Chapter 7

INTERNAL STATE AND GENE EXPRESSION
Transcription and its control

A *gene* is a segment of the long thin molecule of DNA. Each gene encodes a protein or part of a protein. The proteins perform all the numerous activities of life. They are directly or indirectly responsible for all the cell's structures, its organisation, its metabolism, its transport processes, and the co-ordination of these: in short, its internal state. Genes themselves do nothing except encode proteins. To make any given protein, the appropriate gene is copied on to a messenger RNA and then this copy is read by ribosomes, which translate the coded instructions. When a cell reproduces, the replicating machinery has to ensure that each of the two daughter cells receives an exactly identical copy of the parent cell's DNA. Both daughter cells must have all and only the same genes as the parent cell so that they are potentially capable of making all (and only) the same proteins as the parent cell.

The phrase "*potentially* capable" is our point of departure for this chapter. The various different cells in a multicellular organism contain (with very few exceptions) exactly the same DNA, the same genes. Cells in the same organism that differ in function and appearance − i.e. have different internal states - necessarily contain different proteins. Therefore, although they *contain* the same genes they *express* different ones. Changes in gene expression alter a cell's complement of proteins and consequently the internal state.

The control of gene expression
In chapter 2 we compared a gene to a master document in a secure library. The first step in making a protein is to "photocopy the document". The

"photocopy" is a messenger RNA molecule[11], which the ribosomes then "read" and translate to make the protein. We now need to examine the "photocopying" process more closely.

The "photocopier" is an enzyme, *RNA polymerase II*[12] ("polymerase" for short), which copies the relevant part of a DNA strand (the gene) base by base to make a faithful replica. This "copying" process is technically known as *transcription* (*trans* = cross; *scriptus* = written). The message written on the DNA is written out again in the form of RNA. The polymerase starts transcribing at the beginning of the gene and stops at the end.

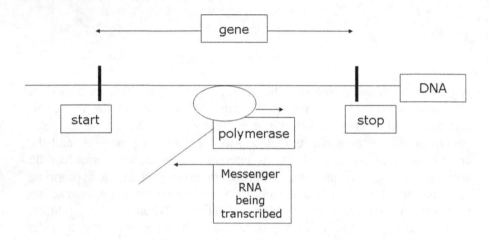

Fig. 7-1(a): an outline scheme of the transcription process.

[11] Strictly speaking, matters are a little more complicated. In prokaryotes, several successive genes are sometimes copied on to a single long messenger. In eukaryotes, the RNA copy of the DNA needs to be *processed* before it becomes a mature messenger; for example, non-coding regions known as *introns* interrupt the sequences that code for the protein, and they have to be cut out of the RNA copy. Important though they are in molecular biology, these matters need not concern us here.

[12] Almost all enzyme names end in *-ase*. In the name of this particular enzyme, the "II" is included because there are other sorts of RNA polymerase, numbered in an arbitrary sequence. The rest of the name indicates that RNA, like DNA, is a *polymer*, a long molecule made by joining together a lot of short molecules (of which the bases A, G, T and C that make up the code in DNA are parts). RNA polymerase joins together some short molecules (containing bases) to form a polymer, RNA, which is a replica of the gene.

b

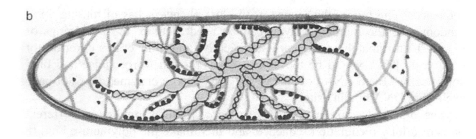

Fig. 7-1 (b): transcription and translation occur simultaneously in a prokaryote. The double-coiled lines represent the DNA, and the lines with rows of dots (ribosomes) are the messenger RNA molecules.

The polymerase runs along the DNA like a toy engine along its tracks, transcribing as it goes. It starts from where it is placed on the rails and stops when it hits the buffers. How is it placed on the rails at the right place, i.e. the start of the gene; and what are the buffers at the end? The answers are, yet again, provided by specialised proteins, which are designed to bind to particular DNA sequences.

Using the standard four-letter code of DNA representing the four bases (A, G, T and C), suppose a short piece of DNA had the sequence

TTGTC**CCAGT**TGGCAAAT**CTTT**T.
       1   2

Consider two DNA-binding proteins. Suppose one binds only to the sequence CCAGT and the other to the sequence CTTT. In the fragment of DNA we have shown here, the former will bind at site 1 and the latter at site 2, but neither protein will bind anywhere else. The sequences CCAGT and CTTT are the *recognition sequences* for these two proteins.

Proteins that bind specifically to DNA sequences at the ends of genes serve as buffers. They stop the polymerase and knock it off the rails. Because the binding is specific, the "buffer" cannot bind to the wrong piece of DNA and jam the polymerase in mid-gene. As for *starting* the transcription process, the simplest design would have the same polymerase recognition sequence at the start of every gene. The polymerase would bind to this sequence, so it would always be placed on the DNA rails in the right place.

This is more or less what happens in prokaryotes. In eukaryotes, however, the situation is a little more complicated. There is so much more DNA in a eukaryotic cell that there is a far greater chance that a 4-5 base recognition sequence, to which the polymerase might bind, will turn up in an inappropriate place. If that happened, the polymerase would waste time and energy transcribing chunks of DNA that are not complete genes. In principle, the solution to this difficulty is to use a longer recognition sequence. The longer the sequence, the less chance it has of turning up at random[13], so the more reliably it can be used to mark the beginnings of genes. Unfortunately, a sequence sufficiently long to meet this criterion for eukaryotic DNA would be too long for any protein to recognise and bind specifically. Even RNA polymerase II, a very large enzyme, does not have such a big DNA binding site.

The practical solution is to have *several* proteins binding to different parts of a long recognition sequence, and then make the polymerase bind to these proteins. The cluster of proteins that binds to this long recognition sequence (the *promoter*) is called the *initiation complex*. Its role is akin to that of a child's hands placing a toy engine on the track at the desired place.

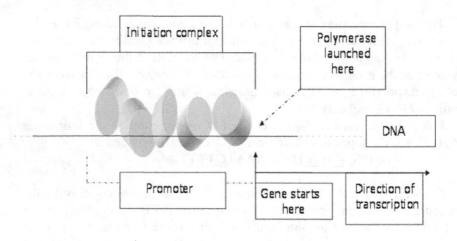

Fig. 7-2: initiation of transcription.

Eukaryotic transcription is started and stopped at the beginning and end of a gene by clusters of proteins bound at the promoter and the termination sites. These clusters cause the polymerase to start and to stop in all and only the right places. But this does not tell us how transcription is *controlled*. Why is a particular gene expressed (transcribed) at some times but not at

---

[13] Since there are four bases, the chances of a particular base turning up in a given position are one in four. The chances of a particular two-base sequence are one in sixteen; of a three-base sequence one in sixty-four; and so on. A five-base sequence has a probability of one in 1024; it is likely to turn up about 6000 times more frequently in human DNA than in a prokaryote. The shortest sequence that is statistically likely to be unique in human DNA is about 17 bases long.

others - switched on and off? And how can transcription be speeded up or slowed down? If there were no practical answers to these questions, there would be no way, for example, of making one type of human cell differentiate from another or adapt to changing needs.

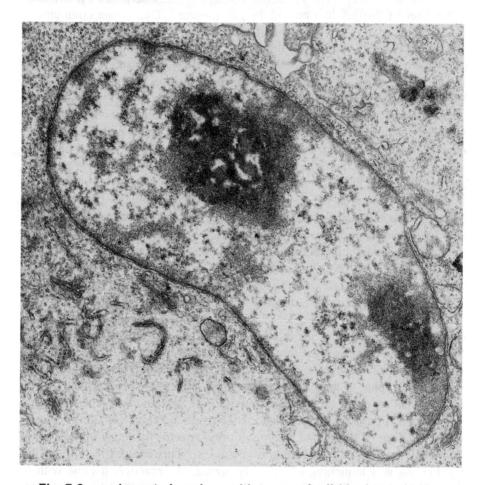

Fig. 7-3: an elongated nucleus with two nucleoli (the large darkly-stained inclusions). The condensed chromatin is visible as smaller darkly-stained areas, while the extended chromatin stains more lightly. The nuclear envelope is also darkly stained.

Genes can be switched off in several ways:-

- The DNA can be modified so that the initiation complex proteins cannot bind, or the polymerase cannot transcribe. DNA that is modified in this way is usually multiply coiled and very compact. This type of modification is often, though not always, irreversible.

- Another way, more readily reversible, is to alter one or more of the initiation complex proteins so that the complex cannot form. If no initiation complex forms, transcription cannot start.
- Yet another way, also readily reversible, is to bind a blocking protein (a *repressor*) near the promoter. This interposes a premature set of buffers in front of the polymerase. This is the commonest method in prokaryotes, but it occurs in eukaryotes as well.

If a gene is not switched off then it will be expressed (= transcribed) - but only slowly. The initiation complex launches a polymerase molecule along the gene once every so often. Slow transcription is not a problem so long as the cell needs only small quantities of the protein encoded in this particular gene. This is the case, for example, for major metabolic pathway enzymes. However, other proteins are needed quickly and in large amounts, and are needed at some times but not others. To meet such needs, the cell must be able to de-suppress the right genes at the right times. Moreover, it must be able to accelerate the transcription of those genes. De-suppression is simple in principle, so long as the gene is switched off reversibly; all the cell needs to do is to reconstruct the initiation complex or remove the repressor. But how can transcription be accelerated? How is the initiation complex persuaded to launch polymerase molecules along the gene faster than usual?

This is done by proteins known as *transcription factors*, which bind to regions of the DNA (*enhancers*) that are often very distant from the gene. This sounds like "action at a distance", or even magic. But remember, to fit the thread representing the DNA into the matchbox model of the cell (chapter 2), you had to tangle it. This tangling might bring two points on the thread a metre or more apart into close contact. Imagine a few grains of salt stuck to one of these two points. Let these grains of salt represent the initiation complex at the promoter (start) of a gene. Now imagine a single grain of salt stuck to the "distant" point. The protein represented by this single grain is the transcription factor attached to the enhancer site. Because of the tangling of the thread, the transcription factor has been brought into immediate contact with the initiation complex. This enables it to speed up the binding and launching of the polymerase. In practice, a gene with a controllable expression rate usually has many enhancers that bind different transcription factors, and their effects are additive. Working together, they accelerate transcription very markedly. When some of them operate and some do not, a more moderate acceleration is achieved[14].

---

[14] A few transcription factors inhibit transcription rather than accelerate it. They are normally outnumbered by the positive factors but they are useful because they make subtle changes in the transcription rate possible.

In short: some genes, such as those for metabolic pathway enzymes, tend to "tick over", transcribing at a constant slow rate. But the outputs of other genes can be varied from zero (when the gene is switched off) to a very high rate (when all the transcription factors on all the enhancers work in concert).

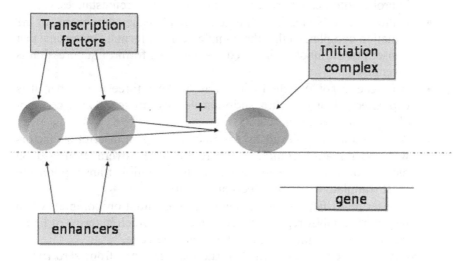

Fig. 7-4: transcription factors, enhancers and the initiation complex.

Controllable genes behave like analogue computers

In this chapter we have presented a lot of new information and several new technical terms, so a summary might be useful at this point.

- A gene is a segment of DNA that codes for a protein. The protein is made when the gene is *expressed*.
- The first (and crucial) step in expressing a gene is to *transcribe* it, i.e. to make an RNA replica or "photocopy" (a messenger).
- Transcription is carried out by an enzyme called *RNA polymerase II* ("polymerase" for short).
- Transcription begins when the polymerase is placed on the DNA at the gene's *promoter* and ends when the polymerase reaches the gene's *termination site*.
- Promoters and termination sites are DNA sequences that bind specific groups of proteins.
- The proteins that bind at the promoter are collectively called the *initiation complex*. It is the initiation complex that launches the polymerase on to the gene; that is, it initiates transcription.
- One way of switching off a gene (preventing expression) is to modify the DNA chemically so that the polymerase cannot function.

Such modification is often irreversible. The modified DNA is usually very compact.

- Another way of switching off a gene is to introduce a repressor protein that binds to the DNA near the promoter and prevents initiation of transcription. This way of suppressing genes, common in prokaryotes, can be reversed under the right circumstances.

- A third way of switching off a gene is to alter the components of the initiation complex so that the complex cannot form. This means that transcription cannot be initiated. Again, this form of suppression is reversible.

- If a gene is not switched off in any of these three ways, then it is expressed. However, transcription usually proceeds at a constant *slow* rate unless it is accelerated.

- Acceleration is achieved by the interaction of transcription factors with the initiation complex. There are often several transcription factors for any one gene. Some of them inhibit transcription but most stimulate it. Their effects are additive.

- Transcription factors are presented to the initiation complex when they bind to DNA regions called *enhancers*, which in terms of linear measurement might be very distant from the gene.

A controllable gene, expressed at a rate that can vary from zero (when the gene is switched off) to a high maximum (when all the transcription factors are acting in concert), is rather like an analogue computer. Think of the transcription rate, i.e. the rate of production of messenger RNA, as the output. It is continuously variable. Think of the transcription factors as the inputs. Think of the initiation complex as the integrator, summing the inputs and modulating the output accordingly. A eukaryotic cell contains numerous controllable genes - a bank of analogue computers. This analogy might help some readers to picture the control of gene expression more vividly. We shall not develop it further at present, but we shall return to it in the final chapter.

How to get rid of proteins

This account of the control of internal states might seem incomplete. We have made it appear that cells just go on expressing genes, sometimes faster, sometimes slower, but making more and more proteins all the time. This is true; but cells do not choke or burst as a result of overproduction, because neither proteins nor messenger RNA molecules last forever. Indeed, some messengers, and some proteins, have very short life spans - just a minute or two. Proteins can be marked for rapid removal by attachment of a "tag". This tag ensures they are quickly destroyed or otherwise rendered inactive, a necessary precaution for proteins that are needed for some immediate purpose but might be harmful if they lasted longer.

Most eukaryotic proteins and messenger RNAs survive for hours or days rather than minutes; but like all machines, proteins wear out. Worn-out proteins are digested by proteasomes (see chapter 3). The component parts of the digested proteins (amino acids) are recycled. Messenger RNA molecules meet a similar fate. Even the longest-lived proteins and messengers will be removed sooner or later. Thus, proteins are continually being destroyed as well as made, so the cell does not become overloaded.

The removal of proteins alters the internal state, just as protein manufacture does. A cell's internal state depends on its protein composition, and in principle this can be as well be changed by removal as by addition of proteins. However, the control of protein breakdown has not been nearly so intensively studied as the control of gene expression. Overall, it is probably less important for altering a cell's internal state because, so far as we know at present, it is subject to less elaborate controls. It is dangerous to be dogmatic about this; we might have to eat our words in ten years' time. But in chapter 8 we shall offer some justification for our claim that the control of gene expression has a more important role than the control of protein breakdown in regulating and manipulating the cell's internal state.

Chapter 8

SUSTAINING AND CHANGING THE INTERNAL STATE
How gene expression and internal state interact

Gene expression and its control form a large part of modern molecular biology and are directly relevant to our definition of "livingness". By changing the genes that it expresses, a cell alters its internal state. The alteration might be trivial: a mere re-adjustment, returning the cell to a *status quo* that has been transiently upset. But it might be radical, a dramatic change in appearance and behaviour.

Two of the questions arising from chapter 7 now become prominent:-

- If the expression of a gene is controlled by transcription factors, why do these factors work some of the time but not all the time?
- Suppose a gene is switched off, for instance by a repressor. What switches it on again exactly when the cell needs it, no earlier and no later?

There is a single general answer to both questions: proteins, including transcription factors and repressors, can be modified. Modification of a repressor or a transcription factor might prevent DNA binding; alternatively, DNA binding might not be possible *without* modification. Modification of transcription factors and repressors ensures that the cell expresses the right genes at the right times. (Also, protein synthesis can be regulated at levels other than gene transcription: processing of the messenger after transcription; messenger transport from nucleus to cytoplasm; and translation by the ribosomes. Moreover, the life-span of the messenger in the cytoplasm can sometimes be altered. But although these levels of control are significant, we shall ignore them here; it is control of transcription that primarily determines whether a particular protein is made. If we discussed the other mechanisms it would complicate the picture.)

Suppose the cell takes in a large quantity of a particular nutrient, such as glucose. This happens in your liver cells after you have eaten a meal. Enzymes that convert the nutrient into storage form (e.g. glucose into glycogen) have suddenly become necessary. One way to meet this need is to accelerate the transcription of the genes encoding these enzymes. This acceleration is achieved by modifying the regulatory proteins, and the modifier might be the nutrient molecule itself. Fig. 8-1 shows a hypothetical scheme of this kind.

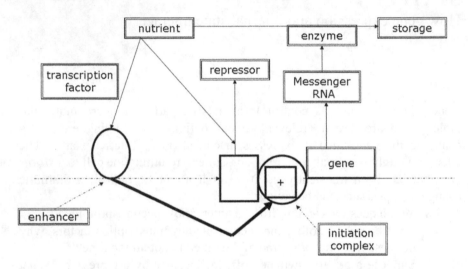

Fig. 8-1: illustration of how gene expression can be controlled. In this hypothetical example, nutrient molecules increase the rate of production of an enzyme that puts the nutrient into storage.

In this scheme, a nutrient molecule binds the repressor, causing it to detach from the DNA. This allows the gene for the storage enzyme to be transcribed. Another nutrient molecule binds to a transcription factor, enabling it to bind to its enhancer. The bound transcription factor activates the initiation complex, accelerating transcription. Rapid messenger RNA production ensues, the requisite enzyme is made, and many millions of nutrient molecules are put into storage as required. When nearly all the nutrient has been converted to storage form, its level in the cell is consequently much lower. So there is no longer enough free nutrient to bind the transcription factor or the repressor. The transcription factor therefore becomes detached from the enhancer and ceases to function. The repressor binds to the promoter again. Transcription stops; production of the messenger RNA for the enzyme ceases.

This hypothetical example illustrates how the mechanisms we discussed in chapter 7 can be used to turn gene expression on and off. However, mechanisms of this kind only *preserve* the cell's internal state. They ensure that the cell's overall behaviour is more or less unchanged despite a large perturbation, such as a sudden influx of nutrient molecules. This extends the idea of *homeostasis* that we introduced in chapter 6. Changes in gene expression that resist changes in the internal state are fundamental to cellular homeostasis.

The time factor
However, there is an important difference between the cellular homeostatic mechanisms we reviewed in chapter 6 (compartmentalisation and enzyme control) and the control of gene expression. The mechanisms outlined in chapter 6 are usually very rapid. Typically, they have time-courses in the order of milliseconds. But a change in gene expression takes effect much more slowly: the new protein appears in minutes, not milliseconds. This is why we regard the homeostatic mechanisms of chapter 6 as *aspects of internal state* and the control of gene expression as *a way of adjusting or altering the internal state.*

Imagine a series of times: t_1, t_2, t_3 and so on. At each of these times the cell has a particular internal state, S, and a particular pattern of gene expression, G. "G" represents the set of genes that are being transcribed and their transcription rates. Let us use "S_1" to mean the internal state at time t_1 and "G_1" to mean the pattern of gene expression at time t_1. In the nutrient-storage case outlined above, S_1 includes the suddenly increased amount of nutrient, and G_1 includes the rapid expression of the "storage enzyme" gene. It will be several minutes before this enzyme is available to the cell; we have now reached time t_2. Once the enzyme is available, the internal state is changed: the nutrient is converted to storage form and the amount of free nutrient in the cell falls. S_1 becomes S_2.

In so far as S_2 is different from S_1, it will alter the pattern of gene expression, which now becomes G_2. But it will be time t_3 before G_2 affects the internal state, changing it from S_2 to S_3. The new internal state S_3 will then change G_2 to G_3. And so on.

The cell does not really make a series of sudden jumps (from G_2 and S_2 at t_2 to G_3 and S_3 at t_3 etc.). There is a succession of smooth changes in S (internal state) and G (pattern of gene expression) over a continuous time-course. Genes differ in their expression rates and their responses to transcription factors and repressors. They are not all switched on and off,

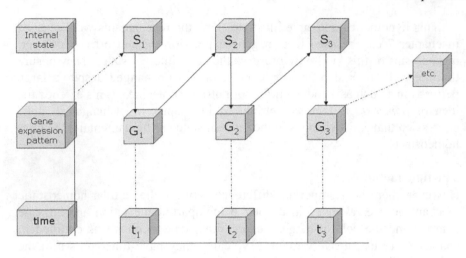

**Fig. 8-2: gene expression and internal state influence each other, but in
one direction there is a time delay.**

accelerated or repressed at the same time. Nor are their transcription rates
all identical. Nevertheless the message of Fig. 8-2 is valid. The internal
state affects the pattern of gene expression more or less immediately; but
changes in gene expression alter the internal state after a delay.

Suppose the cell needs to maintain its internal state, not to change it.
(This is the situation we have been considering up to now.) Events such as a
sudden uptake of nutrient perturb the internal state (say at time t_1). This
perturbation lasts until the consequent changes in gene expression take effect
(time t_2). But once the gene expression pattern has been changed, the
change is likely to persist for a while. A gene switched on at time t_1 might
be switched off again at t_2, but the enzyme or other protein made while the
gene is active might not be removed or inactivated immediately. Indeed, the
messenger RNA for this protein might be stable. This can perturb the
internal state in a different direction; the cell "overcorrects". Gene
expression then changes once more to correct the overcorrection. This
oscillating behaviour can go on more or less indefinitely. In some respects,
a cell's internal state tends to behave like a car fishtailing along an icy road.
Every time the rear of the car goes out of line the steering is adjusted; the car
overcorrects; the steering is adjusted again; the car overcorrects again; and
so on. A cell's internal state tends to oscillate over time.

The effects of this are apparent in some hormone-secreting cells. When
they are active, these cells do not usually secrete the hormone smoothly and

continuously, but in a succession of short pulses[15]. The blood stream irons out these pulses so that the target tissue experiences a steadily increased concentration of the hormone, but the behaviour of the secreting cell itself is oscillatory.

Reproduction, growth, differentiation and gene expression

As we have seen, the delayed response of gene expression to internal state can result in oscillations around median values. This is the situation when the cell needs to maintain a "constant" internal state. But the same underlying phenomenon, the delayed responses of genes, can also lead to *progressive changes* in internal state and gene expression pattern. Transcription factors are proteins; they too are products of particular genes. A single transcription factor can influence the expression of several genes, including those encoding other transcription factors. This is the principle underlying progressive changes of internal state.

Suppose a transcription factor (let us call it F_1) is present when the internal state is S_1. Suppose F_1 causes genes A, B and C to be expressed, and that gene C encodes a second transcription factor, F_2. When the proteins encoded in genes A, B and C have been made, the internal state is S_2 - different from S_1. In particular, F_2 is now present. Now suppose that F_2 causes genes D, E and H to be expressed, changing the internal state to S_3. If H (say) encodes yet another transcription factor, F_3, then a progressive change is underway. Thus, when the pattern of gene expression involves genes for transcription factors, changes in gene expression change the internal state progressively. S_3 succeeds S_2 as G_2 succeeds G_1. Such progressive change goes in a predetermined direction in predetermined stages: it is *programmed*.

Programmed changes in cells can take several different forms. They can be cyclic, so that after a succession of internal states S_1, S_2, S_3 ... the cell returns to S_1. (There are corresponding changes in the gene expression pattern: G_1, G_2, G_3 ... and back to G_1.) Eukaryotic cell division involves a complicated process known as the *cell cycle*, which exemplifies this kind of progression. At one internal state during the cell cycle, the cell's entire

---

[15] In some cases this pulsatile behaviour depends indirectly on the overcorrection phenomenon described in the text; in other cases it might have a different though analogous cause (reciprocal influences of one cell type on another). This is a topic for the next chapter, where we discuss the responses of cells to stimuli from outside, including signals from other cells.

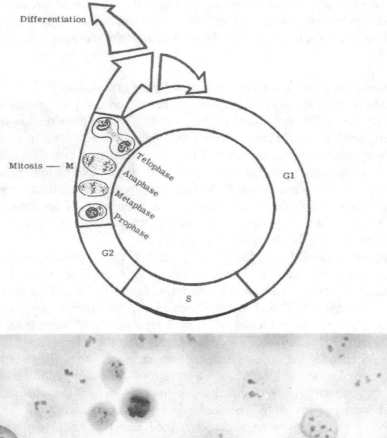

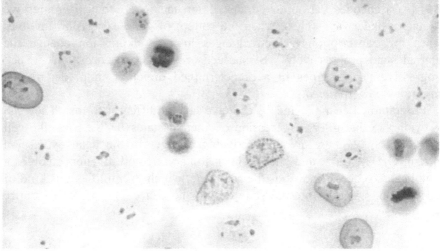

Fig. 8-3: the cell cycle. The diagram (upper part of the picture) shows the division of the cell cycle into the division phase (mitosis) and two growth phases (G1 and G2) separated by the phase of DNA replication (S). The lower picture is a micrograph of a population of cells at different stages in the cycle; some are dividing (the darkly stained contents are condensed chromosomes).

DNA is copied (duplicated). In a subsequent internal state, the daughter chromosomes condense and separate. This is followed by division of the cell into two daughter cells, both of which return to the initial state (S_1) of the parent cell. The cell cycle then commences again. One turn of the cycle doubles the number of the cells, so this process is fundamental to both *reproduction* and *growth* in eukaryotes. Thus, another two of the seven traditional properties of living organisms emerge from the interplay between gene expression and internal state.

Another form of programmed change is *cell differentiation*. In chapter 3 we mentioned the two hundred different types of cells making up a human body. This variety is the result of differentiation. Differentiation involves a linear sequence of internal states and patterns of gene expression, not a cyclic one. The progression is finite; there is a final state known as *terminal differentiation*. In a differentiating cell, a progressively smaller and smaller set of genes is expressed at a progressively higher and higher rate. A terminally differentiated cell might have very high rates of transcription of half a dozen genes, perhaps even just one or two. Genes that are required for basic metabolism and structural maintenance are also expressed, but usually at low tick-over rates. More or less all other genes are switched off. Energy and manufacturing resources are focused almost exclusively on developing the activities necessary for the cell's specialist role in the body. Amongst the genes that are switched off during differentiation are those encoding the cell cycle proteins; terminally differentiated cells usually cannot divide.

A third form of programmed change is programmed cell death or *apoptosis*. In multicellular organisms, almost any cell type seems capable of embarking on a sequence of gene-expression and internal state changes that is ultimately fatal. This is a highly organised form of suicide. No cell contents or debris leak into the rest of the body. Instead, the remains of the dead cell are packaged into small membrane-bound bundles. These are easily engulfed as endocytic vesicles by other cells and digested by their lysosomes (see chapter 3).

At first sight apoptosis might seem a peculiar, negative process, but it is essential for multicellular organisms. In a developing human embryo, for example, the little buds of tissue that will ultimately become arms have flat, blunt ends. To make these flat blunt ends into fingers and thumbs, the cells *between* the incipient fingers and thumbs must be eliminated. This is done

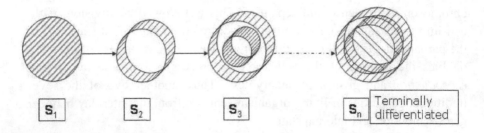

Fig. 8-4: a general scheme of cell differentiation

by apoptosis. Were it not for apoptosis during development, none of us would be born with separate digits. Programmed cell death is also used for removing abnormal or virally infected cells.

The role of protein breakdown
At the end of chapter 7 we said that although protein breakdown is a controllable process, it is probably less important than protein synthesis (gene expression) for regulating the internal state. We can now offer a plausible justification for this remark. Compared to transcription, protein breakdown is usually rapid. To inactivate a protein and to prepare it for disposal is usually the work of seconds rather than minutes. Therefore, while gene expression is best regarded as programming for the *next* internal state, protein breakdown is best regarded as an aspect of the *current* internal state.

Nevertheless there is a connection between gene expression and protein breakdown. The destruction and removal of a transcription factor or repressor (not to mention its messenger RNA) can contribute significantly to the pattern of gene expression. Faster breakdown means less of the protein. Every protein molecule has a finite lifespan, beginning with the initiation of transcription of the gene and ending with inactivation and dissolution, so its concentration in the cell can be altered by changing either the production rate or the removal rate. Protein breakdown receives little attention from present-day molecular biologists. Its true importance will only become clear after more research.

Some transcription factors have long lifespans but are only briefly active. Most of the time they are bound to immobile structures in the cytoplasm. In order to reach the nucleus and bind to their enhancer sequences, they have to be liberated from these restraints.   Only certain particular internal-state conditions allow this to happen. So it is only when these conditions are met that the transcription factors function, switching on the genes to which they are related.

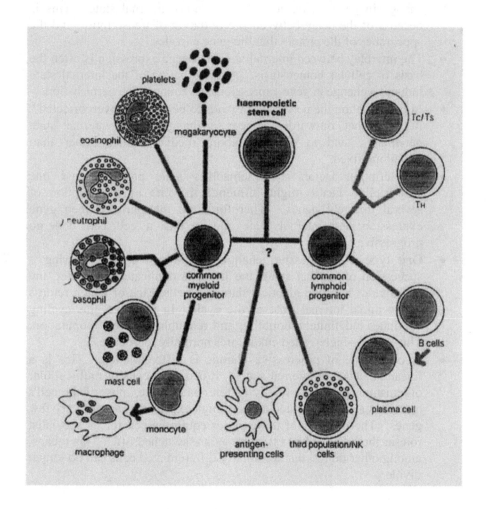

Fig. 8-5: showing how a wide variety of mammalian blood cells differentiate from common precursors.

Summary

- Gene expression and internal state are mutually dependent and influence each other in a variety of ways. Once again, the familiar pattern of reciprocal dependence has appeared. Crucially, however, while gene expression is influenced by the *current* internal state, a change in gene expression alters a *future* internal state. This is because of the delay between the initiation of transcription and the appearance of the protein that the gene encodes.

- The interplay between internal state and gene expression is often the basis of cellular homeostasis. A perturbation of the internal state induces a change in gene expression that counters the perturbation.

- In this situation the perturbation tends to be "serially overcorrected". In cells that outwardly appear unchanging, the internal state parameters tend to oscillate about median values rather than remaining fixed.

- Transcription factors are themselves gene products, and one transcription factor might influence the rate of transcription of several different genes. Therefore, the interplay between gene expression and internal state means that a cell can undergo progressive change.

- One type of progressive change is the cell cycle. During a succession of internal states the DNA is duplicated and, later, the cell divides into two identical daughter cells, each of which returns to the initial internal state of the cycle. In principle, this cycling continues indefinitely, doubling and redoubling the cell population. This is how single-celled eukaryotes normally reproduce.

- Another type of progressive change is differentiation. This is a linear sequence of internal states. It occurs in cells of multicellular organisms. In the terminally differentiated state, most of the cell's resources are concentrated on very high expression rates of very few genes. The products of these genes equip the cell for its specialist role in the body. Most other genes are switched off. This means, among other things, that terminally differentiated cells can no longer divide.

- A third type of progressive change is apoptosis. Again, this is a linear sequence of internal states, but it leads to the death of the cell. The remnants of the dead cell are packaged in small vesicles; these are engulfed by other cells, digested by their lysosomes and recycled. Apoptosis is necessary for removing superfluous cells, for instance during development, and for destroying infected or otherwise damaged cells.

Chapter 9

RESPONDING TO THE ENVIRONMENT
Signal processing, gene expression and internal state

So far, we have touched on six of the traditional properties of living organisms: eating, breathing, excreting, moving, growing and reproducing. The only one that remains is "responding to outside influences". A dog salivating at the smell of food, a flower opening in sunlight and a worm crawling towards moisture are examples of organisms responding to *stimuli* from their surroundings. All organisms respond to their surroundings in order to improve their chances of survival and reproduction.

Single cells also react to their environments, detecting signals and responding appropriately. Thanks to many years of research in cell and molecular biology we now understand – in considerable detail in some cases - how they do this. The principles are quite simple, though (as ever in biology) the details are complicated. We shall concentrate on the principles.

If a single-celled organism moves from place to place by swimming, it will swim towards food. It will also avoid noxious stimuli; for instance, it might swim from more acidic to less acidic water. The direction in which the organism swims is determined by a two-stage process. First, the organism must *locate* the food or *measure* the acidity of the water; that is, it must *detect* the relevant stimulus. Second, it must then use its swimming apparatus to move in the right direction - *respond* to the stimulus. Of course, the response must be precisely related to the stimulus. There must be a "connecting mechanism" between stimulus and response. If this connection were not precise, either the response would be inappropriate or there would be no response at all.

In multicellular organisms, stimulus and response are directed towards the wellbeing of the whole organism rather than the individual cell. For example, a cell that secretes a hormone (a) "knows" how much of the hormone the rest of the body needs from moment to moment and (b) makes appropriate adjustments in the amount that it produces. In this case, the stimulus is a measure of the body's need for the hormone, and the response is hormone production. The cell must ensure that the stimulus evokes the response only when required.

The cell that secretes a hormone helps to keep the whole body alive. It benefits from its altruism because it is part of the body; it would not survive if the body died. This might seem different the single-celled organism, where the stimulus (food or acidic surroundings) evokes a response (movement) that is *directly* relevant to the individual cell's survival. Nevertheless the underlying pattern is the same: specific stimulus, appropriate response, and a precise mechanism linking the two.

Almost all living cells respond to a wide range of stimuli. It makes no difference whether the cell is autonomous and free-living, or part of a large multicellular organism. Eukaryotic or prokaryotic, cells respond in precisely-engineered ways to stimuli that are pertinent to their needs. The stimuli might be physical (light, temperature, mechanical contact, etc.) or chemical (nutrient, toxin, specific signalling molecule from another cell, etc.). Broadly speaking, cellular responses to all stimuli, physical or chemical, follow the same basic principles; we shall focus mainly on chemical signals in this chapter.

Connecting stimulus to response

The cell components that detect stimuli are called *receptors*. A receptor consists of one or more proteins and is usually located on the outer face of the surface membrane. In the case of a chemical signal, the signal molecule binds tightly and specifically to the receptor. The general term for a molecule that binds specifically to a receptor is a *ligand*. The receptor undergoes a subtle change of shape when it binds a ligand or is activated by a physical stimulus such as light.

This change of shape alters the receptor's interactions with molecules inside the cell, which are also changed as a result. These intracellular molecules constitute a *signalling pathway*. In effect, the receptor *transduces* the extracellular signal (the ligand or physical stimulus) to an intracellular one, which is conducted via the signalling pathway components.

The intracellular signal modifies the cell's structure and function. It might change the cell's ability to respond to further external stimuli; it might alter the internal state; or it might alter the pattern of gene expression. These alternatives are not mutually exclusive. A single intracellular signal might modify the cell in all three of these ways.

If the receptor is symbolised by **R**, the ligand by **s**, the subtly altered receptor as **R*** and the intracellular signalling system by **M**, we can write this chain of events[16] as

$$R + s \; Rs \; \rightleftharpoons \; R^*s$$

$$R^*s + M \rightleftharpoons R^*sM \longrightarrow R^*s + M^*$$

M* represents the activated intracellular signal - usually a succession of chemical processes rather than a single changed molecule. M* directly or indirectly alters the cell's behaviour. The double arrows in the scheme indicate that each step is reversible; a ligand-receptor (Rs) complex, for example, can split apart into its components, R and s.

As a broad generalisation, therefore, stimulus is connected to response by a three-stage process: receptor activation (e.g. ligand binding); formation of an intracellular signal; and modification of cell behaviour by the intracellular signal.

Signalling pathways and cellular logic

Just as one metabolite is converted to another by a chain of chemical reactions called a metabolic pathway (chapter 4), so a stimulus is connected to cellular responses by a signalling pathway. A diagram showing all the cell's metabolic pathways would be enormously complicated; a diagram showing all its signalling pathways would be at least as complicated. Just as metabolic pathways branch and converge, so do signalling pathways. The cell's information circuitry has many interconnections; signalling pathways "talk" to one another incessantly, just as metabolic pathways do. Metabolic pathways contain feedback loops, where end-products stimulate or inhibit earlier reactions; signalling pathways also contain positive and negative feedback loops. So there is a close formal analogy between metabolic and signalling pathways.

The difference is one of function. Metabolic pathways make energy available to the cell or manufacture the cell's molecular components; signalling pathways convey information.

To illustrate the complexity of cellular signalling, we shall consider two hypothetical and very simple signalling pathways (Fig. 9-1). Each begins with a receptor (R_1 or R_2) being activated by a ligand or physical stimulus (s_1

---

[16] It is difficult to fit a single general model to all cases. For example, some receptor molecules protrude through the membrane. When the stimulus molecule binds to the exterior portion of such a molecule, the interior portion is changed. In such a case, R and M represent parts of the *same* molecule. However, the scheme shown in the text conveys the general idea.

or s_2). Receptor activation transduces the signal to the respective intracellular molecules, M_1 and M_2. Each of these molecules initiates a branched signalling pathway: M_1 via A, B, and two targets of B - C_1 and C_2; M_2 via X, Y, and two targets of Y - Z_1 and Z_2. The first pathway leads via C_2 to response $RESP_1$; the second leads via Z_2 to response $RESP_2$. $RESP_1$ and $RESP_2$ might be changes in one or more aspects of internal state, or gene expression, or membrane behaviour. We have to bear in mind that *many* signalling pathways operate simultaneously in a cell, not just two.

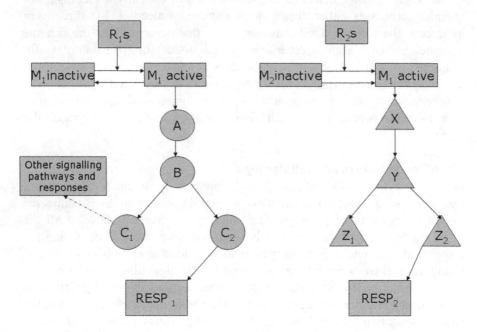

Fig. 9-1: outline scheme of a hypothetical signal pathway.

Now let us consider how these two pathways might interact. Suppose stimulus s_1 inhibits response $RESP_2$, while s_2 inhibits response $RESP_1$. Fig. 9-2 shows just *some* of the ways in which s_2 might inhibit the s_1 signalling pathway. There are just as many ways in which s_1 can interfere with the s_2 pathway, but we have omitted these – they would have made the diagram unreadable.

Even in this highly simplified scheme, you can see how complex the cross-talk among signalling pathways can be. In real life, a cell responds to many different stimuli; each stimulus might evoke several responses and inhibit several others; and many signalling pathways are considerably longer and more extensively branched than we have suggested in the diagrams. Moreover, we have not shown any feedback loops in these schemes. It is little wonder that the study of cell signalling is a very active and challenging area of research in cell biology today.

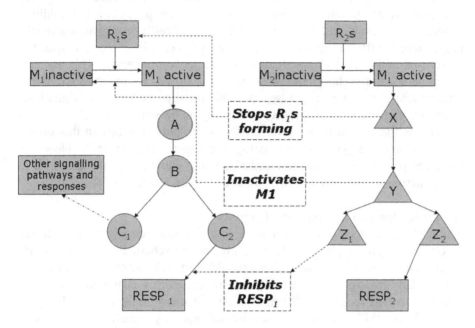

Fig. 9-2: cross-talk between two hypothetical signal pathways. In the interests of clarity, feedback in only one direction (from pathway 2 to pathway 1) is shown.

Several authors have noted an analogy with electronic engineering. The signalling pathway intermediates (M_1, A, B, $C_{1/2}$ and M_2, X, Y, $Z_{1/2}$ in the diagrams) behave like computer logic gates (AND, OR, NOR, etc.). The arrows in the diagram show the ways in which these gates might be interconnected. Cells contain thousands of different sorts of proteins and as many as 25-50% of these might act as signalling pathway components. Such a huge array of logic components, with multiple inputs and outputs, looks like a recipe for chaos. In fact, the system resembles a neural network comprising many interconnected parallel processing pathways; and like a neural network, the cell's signalling system has relatively few stable states despite its complexity. It is not even necessary for all individual components to function precisely. There are so many parallel pathways, i.e. there is so much *redundancy* in the system, that the components can compensate for one another. The response system *as a whole* functions precisely. Stuart Kauffman showed that a network as cross-connected as the signalling system of a living cell is likely to generate a small number of stable states rather than chaos. Interestingly, Kauffman claims that the number of attractors (stable states) of any such network is similar to the number of distinct cell types in the organism.

A system of this kind is capable of *learning*. Exposure to a given combination of stimuli activates[17] some signalling pathways and inhibits others. As a result, responses appropriate to the cell's or organism's needs are evoked. After the response has begun, the cell still contains a specific pattern of activated and inactivated signalling components. This pattern might persist in the short term. If some of the stimuli are repeated while the pattern lasts, the same responses will be evoked. Therefore, signalling pathways confer a kind of short-term memory on the cell.

This is the second time we have used a computer analogy in this book. We compared the gene with an analogue computer in chapter 7. Now we have compared the stimulus-induced signalling pathways of the cell with a neural network. We shall return to these analogies in chapter 18.

Amplification and attenuation of signals

The analogy between cellular signalling and electronic circuitry breaks down in one important respect. If A, B, C etc. in the schematic diagrams were electronic circuit components, there would normally be only one of each. But if they are signalling pathway intermediates this is not the case. A single activated receptor will usually activate many molecules of the type "M". Each "M" will activate several of type "A", each of which will activate several of type "B", and so on. Thus, a very tiny stimulus - the activation of a mere handful of receptors on the cell surface – can induce a very large response in a very short time.

Thus, cell signalling pathways generally *amplify* the external signals. The potential advantage can be seen, for example, in the effect of adrenaline on muscle cells. The leg muscles of a peacefully grazing herbivore use little energy, but the sudden appearance of a predator changes this situation swiftly and radically – this is obviously a matter of survival. The stimulus received by the muscle cells is a tiny increase in the adrenaline concentration in the blood stream. The response includes a huge, immediate increase in the rate of glycogen breakdown and the resulting production of ATP to fuel

---

[17] We have used the words "activated" and "inactivated" throughout this chapter without explaining how a component might *become* activated. Although these details do not matter for the argument in the text, some readers might be curious. In most cases activation (more rarely, inactivation) is caused the addition of one or more phosphates to the protein. The phosphate is usually transferred from ATP. The activation is reversed when the phosphate is removed again. Thus, many signalling pathway components are, in effect, enzymes that cause phosphates to be added to or removed from one another. In other cases, where phosphate transfer is not involved, activation might be caused by the binding of the signalling protein to another protein or to a small metabolite molecule; or alternatively, its *dissociation* from such a molecule.

muscle contraction. The signalling pathway amplifies the glycogen-breakdown response to the adrenaline stimulus.

A cell might have many receptors for each stimulus. For example, a liver cell has a number of insulin receptors on the membrane surface in contact with the blood stream. The more insulin there is in the blood, the more of these receptors become occupied, so the greater the activation of the cell. Therefore, the cell's response to insulin rises with increasing concentrations of insulin in the blood stream. This sensible relationship holds for nearly all stimuli. As the dose of ligand increases, the cell's response increases until all the receptors are occupied.

However desirable the rapid amplified response to a signal might be, an "off switch" is also needed. It is seldom appropriate for the cell to go on responding to a single brief stimulus. There is no single "off switch", but a series of them. The ligand is often destroyed or otherwise removed, thus unloading the receptor; the unloaded receptor is converted back from the active R* to the inactive R form (or destroyed). Activated molecules (M) inside the cell are de-activated. Antagonistic signals are often brought into play, as we showed in Fig. 9-2. Negative feedback from later steps in the pathway can be used to inhibit earlier ones. And so on. These safeguards ensure that the response lasts long enough to bring about the requisite changes in the cell, but no longer. "Long enough" is typically seconds or a fraction of a second.

If the stimulus becomes excessive or abnormally prolonged in spite of these "off switches", the cell might adapt by eliminating some of its receptors. Unwanted receptors are sometimes detached from the membrane and dumped into the environment. In others cases they are pulled into the cell and digested by the lysosomes. In still others they are chemically inactivated. Whatever the method, *receptor downgrading* moderates the response. The cell *adapts*, i.e. becomes less responsive. This stops the cell "burning out" by sustaining its response to a pathologically prolonged stimulus.

Types of response
There are broadly three kinds of response to stimuli: cell membrane changes, alterations in internal state, and alterations in the gene expression pattern.

Two kinds of change might be induced at the membrane level. First, the speed with which something moves into or out of the cell can be altered. In this case the "contact molecule", M, is part of the membrane. When M is activated a specific "gate" is opened or closed, or a "pump" is switched on or off. For example, the activated insulin receptor increases the rate at which glucose enters the cell. Neurotransmitters, the chemicals released from nerve cell termini, alter the rates at which sodium and potassium ions pass through the membrane of the "receiving" cell; this changes the likelihood of electrical activity in that cell. Without the effect of insulin on glucose permeation, the cells of your body would be starved of nutrient. Without the effects of neurotransmitters on sodium and potassium permeation, your nervous system would not work.

Second, another signalling pathway might be modified. The response to one signal might be to inhibit, or to activate, the receptors for another signal. This is another example of cross-talk among signalling pathways.

A subtler "membrane response" is the formation of cell-cell junctions. We mentioned this earlier. The junction might be a tight seal, or it might afford direct communication between the cells (a "zero-resistance junction"). The stimulus for junction formation is direct physical contact between two cells, inducing internal state changes in both. These changes include rearrangements of the cytoskeleton and changes in the patterns of gene expression; so although the final effect is seen at the cell membranes, the response is mediated by changes within the cell.

Many stimuli can affect the internal state without any changes in the membrane other than those directly involved with receptor activation. The "RESP$_1$" and "RESP$_2$" of Fig. 9-2 might, for instance, be the activities of two metabolic enzymes. By making these two enzymes more or less active, the stimuli can dramatically change the relative rates of two or more metabolic pathways. For example, adrenaline makes your heart beat faster and more strongly, while acetylcholine slows it down. Many mechanisms are involved here, but they include alterations in glucose metabolism. The faster your heart beats, the more energy it consumes, so the more fuel it needs. One effect of adrenaline is to activate enzymes of glucose catabolism. One of the effects of acetylcholine is to deactivate them.

A stimulus might also bring about a change in compartmentalisation within the cell. Releasing a substance from a store (or sequestering it in a store) can drastically alter metabolism. When adrenaline binds to heart muscle cells, it causes calcium to be released from internal stores; calcium is not only essential for muscle contraction, it also has further effects on glucose metabolism. The calcium is rapidly returned to storage when the adrenaline stimulus terminates. Alternatively, the stimulus might elicit a change in the cytoskeleton; so the cell might alter its shape or move to a new location, or its internal transport processes might be modified.

In short: via the various branches of its signalling pathway, a stimulus can affect virtually any aspect of the cell's internal state. Indeed, several aspects (metabolism, structure and transport) can be affected simultaneously by the same stimulus.

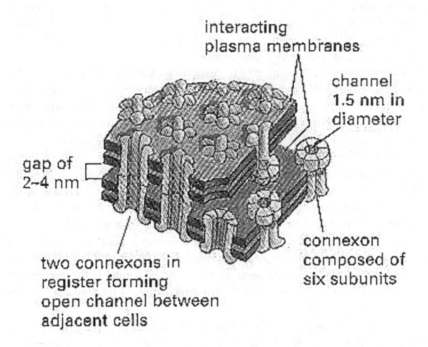

interacting
plasma membranes

channel
1.5 nm in
diameter

gap of
2–4 nm

two connexons in
register forming
open channel between
adjacent cells

connexon
composed of
six subunits

Fig. 9-3: drawing of a cell-cell junction. The membranes of the two linked cells are shown as paired horizontal sheets. The vertical connections between the membranes serve as channels that allow material to be transferred between the cells.

However, the most dramatic and lasting effects of stimuli on cells are attributable to changes in gene expression. A signalling pathway might switch off the expression of certain genes. More commonly, it activates a transcription factor, so one or more genes will be expressed more rapidly. Effects at the gene expression level explain why some cells undergo radical changes in appearance and behaviour in response to external signals.

Responses to stimuli, the internal state, and gene expression
In chapter 8 we described the "dialogue" between a cell's internal state and its pattern of gene expression. We drew attention to the delay between a change in the gene expression pattern and the resulting change in internal state. This combination of dialogue and delay allows the cell to change progressively: the cell cycle, differentiation and apoptosis are possible consequences.

Now we can extend this account. The dialogue is *three*-way, not two-way. In the previous section we outlined the mechanisms by which external stimuli can affect both the internal state and the pattern of gene expression (and the cell's responsiveness to other signals). The converse is also true. Both pattern of gene expression and internal state affect the cell's responsiveness to stimuli.

This is because a cell's capacity to respond to a stimulus depends on the *presence* and the *condition* of the relevant signalling pathway components. If one or more pathway components have not been made, there will be no response to the stimulus. Each component comprises one or more gene products (proteins). So the cell's capacity to respond to a stimulus depends on the pattern of gene expression. If the pattern of gene expression changes, the cell might become responsive to new stimuli or lose its ability to respond to old ones; or it might respond in a different way.

However, each component must not only be present; it must be in the right place and in the right condition. If, because of a particular internal state, a signalling pathway component is chemically modified or locked up in a store and unable to participate in the pathway, the stimulus will elicit an attenuated response or no response at all. So the capacity of the cell to respond to a stimulus depends on the internal state. Once again we find reciprocal dependences:-

- A cell's response to a stimulus depends on the pattern of gene expression; and the stimulus might alter the pattern of gene expression.

- A cell's response to a stimulus depends on the internal state; and the stimulus might alter the internal state.

Internal state affects both responses to stimuli and the pattern of gene expression more or less instantaneously (Fig. 9-4). However, the effects of stimuli are slightly delayed. The internal state at time t_1 affects the responses to stimuli (R_1) and gene expression pattern (G_1) at that time. But the effects of R_1 on internal state (and on pattern of gene expression - not shown in Fig. 9-4) are not seen until later, t_2. The effects of G_1 on internal state (and on the pattern of responses to stimuli - not shown in Fig. 9-4) are not seen until even later, time t_3. It is as though the S directs R and G by telephone, R directs S and G by fax and G directs R and S by snail-mail.

In chapter 10 we shall discuss the relevance of this to our definition or characterisation of "livingness". To finish the present chapter, let us consider the implications of the three-way dialogue for cell differentiation.

Suppose that at time t_0, two cells send signals to each other. By time t_1, these signals have induced alterations in the internal states and patterns of gene expression in both cells. Because of these alterations, both cells might change the signals they send, or they might respond to such signals in different ways. In principle, this is how two cells can direct each other's

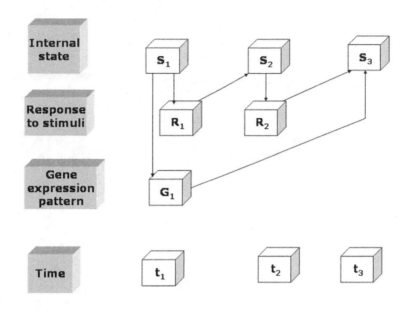

Fig. 9-4: time delays between gene expression, responses to signals and internal state.

differentiation. Cell A might instruct cell B to keep dividing, or make it move to another location; simultaneously, cell A might make cell B anchor itself and over-express certain genes.

In a multicellular organism there are not just two cells, or two kinds of cells, but many. They talk to one another and direct each other's activities. Details of these cell-cell interactions are being uncovered in many areas of research, including embryo development and immunology. Embryo development and the immune system provide dramatic illustrations of the complexity of cell-cell interactions. The three-way dialogue we have outlined in this chapter is manifested less dramatically elsewhere, but it is fundamental to multicellular life. At the whole-body level, homeostasis (chapter 6) depends on cell-cell communications of the sort we have touched on here. Cells respond to signals from one another. Thus, homeostasis is extended from the single cell, as described in chapters 6 and 8, to the vast multicellular assemblies that constitute organisms such as ourselves.

Chapter 10

THE LIVING STATE
A characterisation of 'life'

We suggest that "livingness" is characterised at the cell level by a three-way dialogue among:-
- the internal state;
- the set of all responses to external stimuli; and
- the pattern of gene expression.

"Internal state" encompasses cell structure, metabolism, and internal transport, locked together by reciprocal dependence at any moment. "Responses to stimuli" include all the cell's signalling pathways. "Pattern of gene expression" means the rate (zero or finite) at which *each* gene is being transcribed.

At each of the three corners of the main triangle in Fig. 10-1, details can change from moment to moment. Fluctuations in the internal state affect stimulus-response and gene expression more or less immediately (t_1). Altered responses to stimuli affect internal state and gene expression after a slight delay (t_2). Changes in gene expression pattern affect internal state and stimulus-response after a longer delay (t_3). This dialogue-with-delays can have various consequences:-

- near-constancy, i.e. restoration of the *status quo* after a perturbation;

- a cyclic sequence of changes, e.g. resulting in successive cell divisions (the cell cycle);

- a progressive sequence of changes towards some final cell state (differentiation);

- a progressive sequence of changes resulting in the controlled death and demolition of the cell (apoptosis).

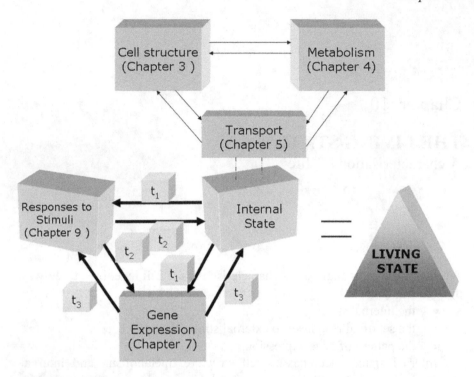

Fig. 10-1: diagram summarising a characterisation of the living state.

In some instances, a change in the internal state might include the production of (new) chemical signals, which are sent out to other cells. It might include the formation of new receptors and signalling pathways. Thus, a cell's capacity to affect and to be affected by other cells can change with time. These changes of capacity underpin embryo development and many other features of multicellular life. On the other hand, the cell's signalling capacities might not change with time, or they might change only in reaction to a disturbance that they seek to correct; thus, homeostasis is maintained.

We have spent several chapters establishing this account of "livingness". Our emphasis has been on eukaryotic cells, particularly the cells of animals such as humans. In this chapter we shall consider two main questions:-

(1) Is the account sufficiently general? Does it apply to plants, fungi and single-celled eukaryotes as it does to animals? Does it apply to prokaryotes? Unless the answers are "yes", the characterisation needs to be reconsidered.

(2) On the other hand, is it sufficiently *restrictive*? Does it apply to anything non-living? If so, it needs to be modified.

Plants

Plant cells have elaborate structures (nucleus, chloroplasts, mitochondria, vacuoles and so on) and a complex web of metabolic pathways. In many plant cells, cytoplasmic fluxes are visible in the light microscope, so there are internal transport processes. Structure, metabolism and transport depend on one another as they do in animal cells. In short, plant cells exhibit *internal states* as defined in chapter 6.

Every plant has its own distinctive genome. Different combinations of genes are expressed in different cells. Changes in internal state, such as increased photosynthesis, increase the transcription of particular genes; the expected dialogue between internal state and gene expression is apparent.

Plant cells respond to external stimuli. Increased sunlight opens pores in the leaves, facilitating exchanges of carbon dioxide and oxygen, compatible with increased photosynthesis. Specific signals promote the growth and differentiation of structures involved in reproduction. Thus, external stimuli can bring about changes in both internal state and gene expression pattern. As in animal cells, the processing of stimulus information requires the manufacture of signalling pathway components (gene products) and their correct location and chemical condition (aspects of internal state).

The three-way dialogue model therefore applies to plants as well as to animals. Our characterisation of "livingness" applies just as well to plant cells as it does to animal cells. However, there are obvious differences between animals and plants. Ecologically, plants are "primary producers"; they manufacture living material out of inorganic substances and external energy. They take water, nitrate and carbon dioxide as nutrients and use a non-living energy source (sunlight). Animals, in contrast, are "consumers"; they obtain their energy and materials from other organisms. They eat either plants or animals that have eaten plants. This basic difference is represented in body design and the overall speeds of life processes: plants are generally less compact than animals and their life processes are slower.

Many animal cells are attached to a framework known as the extracellular matrix, much as sweet peas are attached to a trellis. The extracellular matrix is secreted by certain types of cells and provides support for others, organising their relative dispositions in space and conferring shape on tissues and on the whole body. Plant cells have no extracellular matrix, but unlike animal cells they have tough cell walls. These cell walls can be modified and thickened, lending rigidity to tissues and helping to define shape. The most and familiar example of wall thickening and transformation is the formation of wood. The cells elongate and produce very thick rigid walls with tiny pores opening from cell to cell, then the contents are lost. Wood is a bundle of long narrow thick-walled tubes made of dead elongated cells joined end-to-end by pores. The tubes serve to transport fluids (sap); the thick walls give strength and rigidity[18]. Wood and other rigid tissues fulfil some of the roles that the extracellular matrix fulfils in animals.

Plant reproduction requires the production of spores or seeds. Spores and seeds contain the DNA needed to make a new organism more or less identical to the parent(s)[19]. Seeds (though not spores) usually contain reserve fuel for the plant's growth until it can take care of its own nutrition.

Spores are interesting objects. Encased in tough, resistant coats for protection, they consist of one or a few undifferentiated cells that do nothing. These cells exchange no materials or energy with their surroundings. They do not metabolise. They transcribe no genes, make no proteins. They do not respond to their surroundings unless and until the environment is suitable for a new individual plant to grow. Only then does the spore lose its tough coat and germinate; that is, the cells wake up and become active, and growth and development begin.

Can spores be described as "living"? According to our characterisation they cannot. The cells have internal structures but no metabolism or internal transport processes, so they have no internal states able to change from moment to moment. Moreover, they express no genes and respond to no external stimuli until germination begins, i.e. until the spore ceases to be a spore. If there is no internal state, no gene expression and no stimulus response, then spores are not alive. But they are *potentially* alive, the potential residing in their cellular organisation and a full complement of genes that constitutes a "recipe for making an organism". Spores and seeds are in states of "suspended animation" that can last almost indefinitely. There are well-authenticated instances of spores germinating after thousands of years. Samuel Pepys records the appearance of mustard fields on sites cleared by the Great Fire of London, sites where mustard had not grown since the Roman occupation of Britain.

---

[18] The "grain" and the mechanical strength of timber depend on the size of the tubes and the thickness of the walls.

[19] Some spores (and all seeds) are the results of sexual reproduction, in which case there are *two* parents. Others are produced asexually and are genetically identical to their *one* parent.

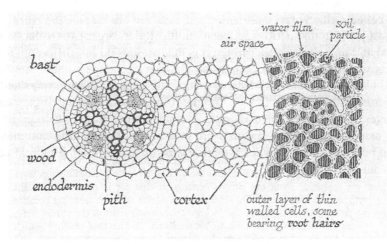

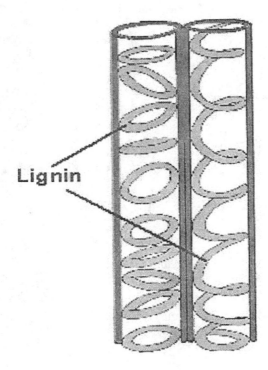

Fig. 10-2: drawings of xylem tubes (wood) in a plant. Upper drawing: cross section. Lower drawing: longitudinal section.

Single-celled eukaryotes
The most salient distinction between multicellular organisms (plants and animals) and unicellular ones (protists) is that protists do not differentiate or, as far as we know, undergo apoptosis. Protists exchange signals, but they respond to these signals only by migration, altered cell division rates and in some cases colony formation. On the other hand, many single-celled eukaryotes have extremely elaborate internal architectures. Compared to animal or plant cells, some protists are of a size and structural complexity that can be quite startling when we are accustomed to studying animal cells such as our own.

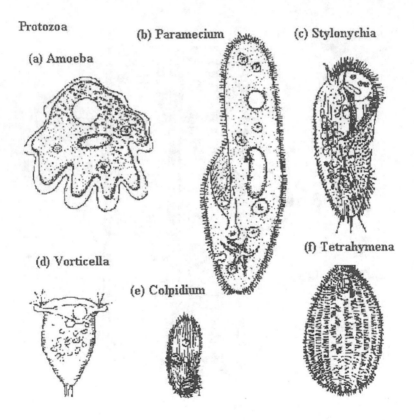

Fig. 10-3: a variety of protists. Drawings of micrographs, not to the same scale.

In all protists, as in plants and animals, there is an internal state (as defined in chapter 6) locked in reciprocal dependence with gene expression and stimulus processing. Our characterisation of "livingness" applies to amoebae, yeasts and other single-celled eukaryotes just as it does to animal and plant cells.

Some protists produce spores or spore-like ("encysted") forms when the environment becomes hostile. Like plant spores, these are instances of "suspended animation"; temporarily at least, they are not living. Their "livingness" is restored when the environment becomes friendly again. Encystment can serve as a method of reproduction among protists, but it is often used simply to cope with hostile conditions.

Prokaryotes
We surveyed some differences between prokaryotes and eukaryotes in chapters 2-3. Prokaryotes are much smaller than eukaryotic cells. They have much smaller genomes, far less elaborate structures and smaller repertoires of responses to stimuli. They do not grow and differentiate as animal and plant cells do. They are machines for reproducing as quickly as conditions allow. When conditions do *not* allow, many bacteria turn into quiescent "suspended animation" forms known as endospores.

In short, prokaryotes are simpler than eukaryotes. However, the characterisation of "livingness" still applies. Bacteria respond to external stimuli. Some of these stimuli promote transcription of particular genes, and the equipment for detecting and processing the stimuli obviously consists of gene products. Transcription and protein manufacture are very rapid, so the delay between stimulus reception and synthesis of a new gene product is considerably shorter than it is in eukaryotic cells. Nevertheless there is still a time lapse. The duration of delay might not affect our characterisation, so long as it is finite; we shall consider this claim at the end of the present section.

Prokaryotes have internal states, as defined in chapter 6, though these are simpler than the internal states of eukaryotic cells. They have interconnected metabolic pathways. They regulate their compositions. They have structures including cell membrane and wall and storage granules. They presumably transport materials between locations within the cell. Metabolism is needed to construct and maintain the cell membrane; the cell membrane is needed to organise metabolism; and perturbations of "cellular homeostasis" change metabolism and membrane organisation. "Simpler than eukaryotic cells" does not mean "simple" in the sense that a non-living physical system is simple. Prokaryotic internal states involve hundreds of different proteins, and even a single protein is a complicated piece of equipment.

Responses to stimuli can change the internal state of a prokaryote directly, as in eukaryotes; and by definition, the signalling pathway intermediates are aspects of the internal state. So there is a dialogue between internal state and stimulus-response. Inevitably, there is also a dialogue between internal state and gene expression, though the regulation of gene expression and its dependence on internal state are less sophisticated than we described in chapters 7-8. Also, prokaryotes respond to signals from one another and from other sources, so they "communicate" in broadly similar ways to eukaryotic cells.

In short, our general characterisation of "livingness" applies to prokaryotes. But it is interesting to consider the shorter delays (the intervals between t_1, t_2 and t_3 in Fig. 10-1). In multicellular eukaryotes these time-delays, notably between gene expression and internal state, are valuable because they allow for differentiation and apoptosis; but no such processes occur in prokaryotes. They also allow for the eukaryotic cell cycle; but prokaryotic replication needs no comparably elaborate cell-division apparatus or succession of events. Therefore, prokaryotes probably do not require the same time-delays as eukaryotic cells. We suggest, therefore, that the shorter time delays do not imply a mismatch between our characterisation of "livingness" and the nature of prokaryotes. Rather, they point to something fundamental in the distinction between prokaryotes and eukaryotes.

Fungi

Essentially, a fungus is a collection of long thin tubes known as *hyphae*.[20] "Long" might be a fraction of a millimeter, or many metres; "thin" is typically around 5 µm, i.e. five thousandths of a millimetre. A fungal hypha is not explicitly or obviously partitioned into cells. Many hyphae appear under the microscope as transparent colourless tubes with nuclei dotted along them. There is only the merest hint of a septum in the space between one nucleus and the next.

Part of the reason for this structural plan might be that most fungal cell walls are made of chitin, the tough, resistant covering material found in arthropods, including beetles and crabs. (Anyone who has tried to kill a cockroach by treading on it knows how strong chitin can be.) Chitin is not conducive to cell-cell communication. If a hypha were divided explicitly into cells by chitinous walls, then each cell would be forced to act as an

---

[20] Specialised parts such as reproductive units are produced from these when necessary, but they are not durable. For instance, mushrooms appear overnight, but soon collapse and disintegrate again.

independent unit. The fungus would be not so much an organism as a colony of organisms.

Rather than reiterate now-familiar arguments, we shall simply state here that fungi *do* fit the characterisation of "livingness". The stimuli to which they respond are usually limited to moisture, nutrients, temperature and sometimes light. These stimuli can activate or inactivate genes, affecting e.g. growth, which can be explosively rapid (as in the case of fungal reproductive bodies; mushrooms appear overnight). They can also affect the internal state. For example, the presence of nutrient induces the hypha to secrete digestive enzymes and absorb the digestion products - a striking similarity between fungi and bacteria.

Fungi eat either living or dead organisms or organic waste. They are not primary producers but are either parasites, which inhabit the host that provides the nutrients, or *detritivores*, causing decomposition of dead organisms or excretory products. Detritivores are ecologically indispensable; they are instrumental in recycling resources within ecosystems. Parasitic fungi lack some of the components they need for independent, host-free existence, as all parasites do, but they are alive: they exhibit a continual dialogue among gene expression, stimulus-response and internal state.

Viruses

Viruses contain proteins and a nucleic acid (DNA or RNA), and some viruses are membrane-limited. Also, they interact very specifically with living cells, often replicating themselves with lethal consequences for the host. They subvert the host's molecular machinery and force it to work according to their own instructions. In these respects, viruses resemble organisms and there is little doubt that they are related to organisms, so they are proper subjects for biologists. Nevertheless they are not alive.

Let us examine this claim, which some biologists might consider heterodox. (Others, such as Lynn Margulis, would agree with us.) First, viruses have no metabolism. They have no devices for making energy and materials available. When they infect a host cell their replication depends on the host's manufacturing and energy-supply systems. They have no internal transport processes and do not exhibit cellular homeostasis. Outside the host they are as quiescent as crystals. They have elaborate and distinctive structures; indeed, a virus can be distinguished from other viruses by its structure. But lacking metabolism, "cellular homeostasis" and transport, they have no internal states.

Second, their genomes are minute, comprising perhaps only half a dozen genes. Even the simplest bacteria have genomes more than a hundred times bigger. During infection the viral genes are transcribed in a prescribed order, but this depends on the host's protein synthesising equipment. Also, *all* the viral genes are expressed during infection; *none* is expressed outside the host cell; so there is no "pattern of gene expression" of the kind we discussed in chapters 8-9, susceptible to alteration in response to internal state changes and external stimuli.

Third, the only "stimulus" to which a virus responds is a receptor on the host cell surface. The virus binds to this and is either engulfed by the cell or injects its nucleic acid through the host membrane. There is no semblance of a signalling pathway. There is no processing of stimulus information.

Therefore, a virus has scarcely any of the main characteristics of the living state that we have identified. Viruses are sometimes called "the ultimate parasites". As we said earlier, parasites lack some components that would be needed for life outside the host, and viruses certainly lack the components needed for independent life. But the metaphor is misleading. Tapeworms, parasitic amoebae, malarial parasites and pathogenic bacteria all consist of one or more cells, and these cells have distinctive internal states and patterns of gene expression, though their stimulus-response repertoires might be limited. Parasites are unquestionably organisms. Viruses, on the other hand, do not have internal states, gene expression patterns or stimulus-response systems, so they are not organisms.

Can viruses be dubbed "potentially living", as plant spores are? Some authors take this position, but once again we consider the analogy flawed. A spore contains all the equipment necessary for the intricate choreography of internal state, gene expression and stimulus-response. It only awaits the signal to start the music. There is no such "suspended animation" in the case of a virus; the necessary equipment is not present. It must be supplied by the host.

When a virus infects a cell there are three possible consequences. First, the viral nucleic acid - or the whole infected cell - might be destroyed. From the virus's point of view this is a failure. Second, the host cell machinery might be subverted to replicating the virus. The end result is the destruction of the host cell and the release of several hundred new copies of the virus, each able to infect a new host cell. Third, the viral genome might be incorporated into the host genome. The host cell will continue to live, but it has been irreversibly if subtly altered. Assimilation of viral genes into the host genome has interesting evolutionary implications, which we shall discuss in a later chapter.

Some mammalian diseases known as spongiform encephalopathies have been attributed to the effects of an infectious "rogue protein", a *prion*. Nowadays, the most familiar of the spongiform encephalopathies is BSE, commonly known as "mad cow disease". The oldest known example is scrapie in sheep. These diseases cause slow, progressive destruction of the brain tissue and a deposition of tangled and highly resistant protein fibrils at the sites of damage. The disease is ultimately fatal. According to current beliefs, the prion closely resembles a normal brain protein. It enters the brain and subverts the organisation of this normal protein, thus producing many more copies of itself. If this account is correct, then a prion behaves very much like a virus. Indeed, these diseases were previously known as "slow virus diseases". But no one would suggest that a single protein molecule is "alive".

Crystals, robots and other inanimate objects

Our characterisation of the living state seems to apply to all kingdoms of organisms, though some readers might balk at the exclusion of viruses. Viruses aside, therefore, the characterisation seems sufficiently *general*. But is it also sufficiently *restrictive*?

A crystal of any substance in contact with a saturated solution of the same substance shares many of the properties of organisms. It exchanges material with its surroundings; molecules are exchanged between the crystal surface and the solution. It might therefore be said to "eat" and "excrete". It also grows. And it can replicate itself by stimulating the formation of similar crystals from the solution. Several authors have drawn attention to these similarities between crystals and organisms and have suggested that crystals, which are plainly not alive, emphasise the difficulty of trying to distinguish sharply between the living and the non-living.

However, our characterisation of "livingness" unequivocally excludes crystals. Crystals do not metabolise and do not have internal transport processes (though under some conditions the units of a crystal lattice change places slowly, and apparently randomly). Like viruses, therefore, crystals can have elaborate structures but they have no internal states. They have no genomes and therefore nothing analogous to a pattern of gene expression. They do not respond to stimuli through signalling pathways. Therefore they are not alive and have nothing approaching the complexity[21] of organisms.

Self-regulating objects such as robots are sometimes considered "living". Their electronic circuitry is certainly complex and even their mechanical parts might have much higher information contents than crystals. Robots detect specific stimuli and respond to them in organised and (all being well) appropriate ways. They contain detailed coded information that defines and controls their responses. External stimuli cause specific parts of this information to be expressed in the robot's actions. However, a robot has no

---

[21] A prokaryotic DNA of one million bases contains two million bits of information. (Given that there is a choice of four bases for each position, there are two bits of information per base.) For human DNA, multiply this figure by about 6,000. Crystals are repetitive structures. Even if the unit crystal is very complicated the information content never remotely approaches these figures.

internal state. Its energy supply is not self-regulated; it does not "metabolise". The spatial organisation of its internal parts is fixed. It does not assemble and maintain itself, or exchange materials with its surroundings in order to do so. Therefore, robots are not wholly autonomous; they do not exhibit the three-way dialogue among internal state, pattern of gene expression and stimulus-response that is characteristic of life.

It would be interesting to know whether any non-living entity fits our characterisation of "livingness". We have not been able to identify one. Provisionally, therefore, we conclude that our characterisation is sufficiently restrictive as well as sufficiently general. It applies to all living things but to nothing else.

Why is the cell the fundamental unit of life?
A single cell fits our characterisation. Indeed, we developed the characterisation by reference to single cells rather than multicellular organisms. But consider any *part* of a eukaryotic cell: an isolated nucleus, a mitochondrion, the cell minus its nucleus, or any other permutation. None of these sub-cellular parts can be deemed "living". Take away the mitochondria and you take away most of energy metabolism, so the cell cannot be supplied with ATP; the internal state cannot be maintained. Take away the nucleus and you take away the genes and therefore the pattern of gene expression. Alternatively, consider a fragment of a prokaryote. The fragment can no longer co-ordinate its responses to stimuli, its pattern of gene expression and its internal state, and hence it is not alive. Therefore, although a *cell* can be alive, no *portion* of a cell can be. Deprive a cell of any significant part and the remnant is dead or dying. In other words, the cell is the smallest possible unit of life.

What use is a general characterisation of the living state?
The general characterisation we have developed has a philosophical or semantic benefit: it "defines" the subject matter of biology without circularity. But this is not its only advantage.

We have spoken about "patterns of gene expression" and we have made explicit reference to DNA, proteins and particular parts of cells. However, the living state as we have characterised it does not *have* to involve DNA or proteins. Our account has been abstract enough to avoid reliance on particulars. Anything that performed the same role as DNA could serve the same purpose: act as a "central library" in which the "documents" can be copied in a controllable way. The term "pattern of gene expression" could still be applied. Anything that could do what proteins do (i.e. be responsible for all the structures and all the activities of the cell!) would serve the same role as proteins. We could still talk about the various aspects of internal state and the mechanisms of response to stimuli. We know of no other material that *could* take the place of DNA or of proteins; but our characterisation of the living state could in principle apply to life on other planets *irrespective of whether that life is DNA and protein based.*

Moreover, our account could prove useful for discussing the origin of life on Earth. Conventional approaches to the origin of life are based on the argument about whether nucleic acids or proteins came first. Irrespective of the answer, they presume that increasingly complicated organisation led to the emergence of the first organisms. This might be so, but we can consider a logical alternative: that there were organisms before the first nucleic acids and proteins. A general characterisation of the living state allows us to explore this possibility, even if only to dismiss it.

These are topics for later chapters. What else have we gained so far? Perhaps what you have read in this book so far might enhance your appreciation of the natural world. The oak, the primrose, the beetle and the weasel each consist of countless cells. Each cell contrives, through the interplay of internal state, gene expression pattern and responses to stimuli, to fulfil its ordained part in the proper functioning of the whole organism. Fig. 10-1 applies to every cell in each of these organisms. Myriad micro-organisms, invisible to the naked eye, house similar interplays, all described by the same general scheme. Working together, these organisms make up an ecosystem, exchanging and recycling energy and materials, controlling each other's population sizes.

In chapter 1 we said that deeper insight and understanding increases rather than decreases our awe and wonder at the natural world. This is true for us; we hope it might be true for our readers.

Chapter 11

STABILITY AND CHANGE IN DNA
How genes can be altered

Proteins conduct the processes of life. Each protein is encoded in one "master document", a gene, in the cell's "library", the genome. The genome consists of DNA. No cell can function without the proteins that it requires, so it cannot function without its genes, its DNA. If some of the "master documents" are missing or defective the consequences are potentially serious. Every cell must begin with a very accurate copy of its parent's or parents' DNA. So it is important for the DNA to be stable, to resist change, otherwise inaccuracies will appear.

This is a challenge. Recall our discussions of the shape and size of a DNA molecule (chapter 2) and the control of gene expression (chapters 7-9). The cell's DNA looks fragile and it is continually being transcribed, replicated, packaged and unpackaged. A very long, thin thread that is incessantly being coiled and uncoiled, stretched and compacted, decorated with adhesive proteins and freed of them again, is vulnerable to damage. DNA is continually harassed and battered by the processes of life. Unless they were constantly subjected to repair and maintenance, the cell's "master documents" would become degenerate.

If a gene is altered, the cell's metabolism or internal structure might be changed, or perhaps its transport mechanisms; in short, its internal state. Or there might be a change in the way it responds to stimuli from the environment or controls the expression of other genes. The long term survival of an organism depends on its ability to prevent or forestall such damage, or to buffer itself against the effects of gene modification.

On the other hand, if genes never changed at all, there would be no variation among organisms, no evolution, no diversity of life. The evening in the woodland would be a far less magical experience. Indeed, there would be no woodland and no one to appreciate it. We owe our existence, and we owe the abundance of life around us, to the capacity of DNA for change. In this chapter we shall examine this seeming conflict between the need for DNA to remain stable and the need for it to undergo permanent alterations.

Protection mechanisms

DNA is chemically stable; more stable than RNA[22]. Large genomes could not be made of RNA because they would degenerate far too quickly. Some viruses (HIV is the best-known example) have "genomes" made of RNA rather than DNA, but (a) viral "genomes" are very much smaller than organism genomes, and (b) RNA viruses are notorious for their capacity to mutate rapidly.

Despite its chemical stability, DNA remains vulnerable to damage, particularly from the products of oxygen metabolism and radiation. Cells contain chemical protection systems such as anti-oxidants that eliminate these damaging metabolites before they do too much damage to the DNA. Limited damage can be repaired.

In addition, the genome is attended by a coterie of molecular "maintenance mechanics" that scan the DNA for damage, mend it, and keep the DNA in working order. These "maintenance mechanics" are enzymes. They detect "bumps" in the double helix that result from copying errors, then they carry out surgery: they cut out the incorrect piece of the molecule and substitute the right one. If the maintenance crew fails in its duty, misprints accumulate in the library's master documents and the cell starts to malfunction.

When we age, the cell's chemical protection systems become less effective and the maintenance crews cannot keep pace with the increased work-load. Errors accumulate in the DNA, cells malfunction and sometimes die, cell-cell signalling becomes abnormal, tissues degenerate, and cancers begin. But ageing is not the only way of inflicting more damage than the maintenance crews can handle. The cumulative effect of environmental factors, such as radiation and toxic chemicals, will eventually alter our DNA, causing cancers and other disorders. Interestingly, anti-oxidants provide some protection against the effects of (for instance) radiation damage.

---

[22] Explaining the difference involves some rather complicated organic chemistry, but essentially it involves the pentose sugar unit in the sugar-phosphate backbone of the nucleic acid. In RNA there is a hydroxyl group at the 2' position of the sugar (ribose) and this hydroxyl group can attack the sugar-phosphate bonds, especially in alkaline solution, breaking the "popper-bead necklace" chain. In DNA, the sugar (deoxyribose) lacks this hydroxyl group so there is nothing to attack the bonds. While this difference confers much greater chemical stability on DNA, it also confers chemical activity on RNA; some important enzymes consist of RNA rather than protein.

In summary, genes are protected against damage by:-
- the chemical stability of DNA;
- cellular protection systems against damaging metabolites;
- repair and maintenance assemblies.

Buffering the cell against gene alterations

These protection systems are effective, but they are not perfect. The threat of DNA damage remains; sustained attrition cannot be resisted altogether. So far as we know, organisms have been susceptible to such assaults throughout the history of life, perhaps more intensely in some eras than others, and not all organisms are equally well defended. Therefore, DNA has always been susceptible to *mutation*. A mutation might involve the insertion, deletion or alteration of a single base (a point mutation), or it might affect more of the DNA sequence.

Although a mutation is usually not good news for the cell or the organism, it might not always be bad news; many mutations have no discernible consequences. One reason is that in most eukaryotes (though not in prokaryotes), a good deal of the DNA is "junk". That is to say, it does not encode proteins. In human DNA, for example, genes make up only 3-5% of the total. The more non-coding DNA there is, the lower the chance that a mutagen will affect a gene. Suppose you are in a crowd where someone starts shooting at random. The bigger the crowd, the smaller your personal risk of being hit; there is safety in numbers.

Moreover, a mutation that does occur within a gene might not alter the protein encoded in that gene. The genetic code contains redundancies, so a change in a DNA letter might not change the protein's amino acid sequence. Also, if an amino acid *is* changed, it might be replaced with one that serves equally well, and the protein's function remains unimpaired. For example, the DNA base sequence UUU specifies the amino acid lysine. That is to say, where UUU occurs in the gene, lysine will occur at the corresponding point in the protein. UUC also specifies lysine, so a mutation that converted the third U to C would not alter the protein at all; lysine would still appear in the same place. UCU specifies a different amino acid, arginine (so does UCC), so if the second U were mutated to C, the protein *would* be changed. But arginine and lysine are chemically similar in many ways, so the replacement of one of these amino acids by the other might still leave the protein functional, though slightly changed.

There is another and rather simpler point. In sexually reproducing organisms, each cell has two copies of nearly every gene, one copy from each parent. If one of these copies is defective, the other will probably be normal. So a mutation in one copy of a gene still leaves the organism capable of making the normal protein. This is why the "carriers" of genetic diseases are often free of symptoms: they have one mutant gene and one normal one. Only those offspring who have mutations in both copies of the gene are affected.

The loss of one gene might make the cell non-viable, but the consequences are seldom so extreme. In these days of advanced molecular biology, it is easy to eliminate almost any gene from an embryo, and in most cases the organism develops more or less normally. "Gene knock-out" has become a routine experimental technique. Many different genes can be knocked out of a mouse without it ceasing to be a viable mouse. Of course there are exceptions, where the alteration or loss of a single gene produces a seriously impaired or completely non-viable cell or organism, but despite the number of genetic diseases listed in our medical text books, not all genes seem to be "essential".

This is not really surprising. As we said in chapter 9, complex systems with redundancy are robust; they can function when individual components are missing. So we ought to expect cells to tolerate a certain amount of gene damage or loss; the rest of the system compensates for their absence or malfunction. The defective cell might lack some structure or activity that the normal ("wild-type") cell has, but in most respects it will be the same cell. Relatively few gene products are so essential for the cell's viability that the system cannot compensate for their absence.

In summary, organisms are buffered against mutations by:-

- "junk" DNA;
- the redundancy of the genetic code;
- the chemical similarities between some pairs of amino acids;
- in sexually reproducing organisms, two copies of nearly every gene;
- compensation for the effects of gene loss by the rest of the system (a complex system made robust by redundancy).

Functional rearrangements in DNA

Mutations are passed on to daughter cells. In multicellular organisms, they might be transmitted to the organism's offspring. Over a series of generations they accumulate. The effect is to produce *variation* within the species. Variation is the raw material of evolution and the wellspring of life's diversity. But simple "point mutations" (changes in a single base, altering a single gene at a single point) are not the only means of altering the DNA. There are more dramatic ways, rarer than point mutations but probably much more influential on the course of evolution.

Sometimes a piece of a DNA molecule can be excised and the ends spliced back together. Occasionally the excised piece can be turned round and re-inserted back to front. This often produces nonsense, but not always. Suppose a deranged editor altered the sentence:-

Napoleon was heard to declare <u>that he was able ere he saw Elba</u> *before his final* **battle**

by deleting either the underlined segment, or both the underlined and italicised segments, and splicing the remaining pieces together. The resulting sentences would have quite different meanings:-

Napoleon was heard to declare before his final battle

Napoleon was heard to declare battle.

There are editors who do this sort of thing. But few would be deranged enough to re-insert a letter by letter *inversion* of the excised underlined phrase, producing:-

Napoleon was heard to declare <u>able was eh ere Elba saw eh taht</u> *before his final* **battle,**

though the utterance might throw light on the outcome of Waterloo. The inverted segment is nonsense, but it *almost* makes sense. A few more point mutations in the right places could make it intelligible again.

Deleting a segment of a DNA molecule can produce new genes: intelligible sentences, so to speak, but with meanings different from the original. Inverting and re-inserting the excised fragment can produce something from which sense can be rescued. These are examples *functional rearrangements* of DNA, and they are significant factors in evolution. For instance, suppose the deleted or inverted region contained a gene promoter or the binding site for a repressor. In this case, deletion or inversion would radically alter the transcriptional activity of at least one gene. Such cases are known.

A genome might contain DNA sequences that can be joined in alternative ways by cutting and splicing. To continue the linguistic analogy, consider the sentence:-

She was struck by the fact that the man was climbing the drainpipe in a pair of hobnailed boots.

Omission of different segments of this sentence yields at least three sentences with entirely different meanings. We might describe the procedure as "sentence conversion".

She was struck by the fact that the man was climbing the drainpipe.

The man was climbing the drainpipe in a pair of hobnailed boots.

She was struck by a pair of hobnailed boots.

Segments of DNA can be by-passed in a way analogous to the formation of these three sentences from the "parent" sentence. The process is called *gene conversion*. In some species of yeast, the process generates different "mating types", and these are relevant to the organism's reproduction.

Some genes become *amplified*: multiple copies are made and inserted into the DNA one after the other, generating a series of more or less exact repeats. Amplification is useful or even essential to the cell when very large amounts of a gene product are needed quickly. This is the case with some components of ribosomes. (Recall that ribosomes are the protein-making machines that read the messenger RNA "photocopies". Every living cell needs very large numbers of ribosomes to cope with its protein synthesising requirements.) Other examples include the genes for the eukaryotic DNA packaging proteins, the histones. The total amount of histone required by a cell is similar to the total amount of DNA: a lot. However, histones can only be made at the moment in the cell cycle when the DNA is duplicated; excess, unbound histones would cause terminal damage to other components. Therefore, the histone genes are amplified; there are multiple copies of them. These genes are transcribed rapidly when DNA duplication occurs and switched off again immediately afterwards.

Genes are sometimes amplified when the cell has no need for multiple copies. Since the extra copies are redundant, it makes little difference to the cell if they mutate almost out of recognition. They come under the heading of "junk DNA". Nearly all eukaryotic genomes contain the mutated remnants of unwanted genes, which might be the products of amplification events in the remote evolutionary past, and these become scattered all over the genome, not necessarily adjacent to the site of the original gene. These functionless remnants are called *pseudogenes*. But, on a rare occasion - as a result of serial mutations - pseudogenes acquire new functions, becoming relevant to the cell and to evolution once more. Such "rare occasions" might, in fact, have happened fairly frequently. DNA has been changing for almost four thousand million years since the origin of life, so there has been plenty of opportunity for "rare events".

Occasionally a very short segment of DNA, perhaps only 3 or 4 bases, is copied over and over again, producing *simple-sequence DNA*. One such reiterated sequence makes up about 10% of the human genome. The difference in genome sizes among complicated organisms owes more to the quantity of simple-sequence DNA than it does to the total number of genes.

In summary, some of the more dramatic ways in which DNA can be altered are:-

- excision, splicing and inversion;
- gene rearrangements;
- amplification;
- formation of pseudogenes;
- production of simple sequences.

Transpositions

This list is not complete. Genes occasionally leap from place to place in a genome, even from one chromosome (DNA molecule) to another. This is another "rare event", but probably of great significance in evolution. There are several different mechanisms of *transposition* but the commonest involves the insertion of viral DNA near to the gene of interest. We mentioned this phenomenon in chapter 10; viral infection sometimes results in the viral DNA hiding in the host cell's genome. Often the virus involved in such an event is a *retrovirus* - it contains RNA instead of DNA – so it replicates by making a DNA copy of its RNA ("reverse transcription"). Then the DNA copy hides in the host genome. When this region of DNA is transcribed, part of the resulting RNA resembles the original retrovirus RNA; other parts include transcripts of nearby genes. The entirety of this RNA might then be reverse-transcribed and the resulting DNA inserted into a new place in the genome.

Human DNA, indeed the DNA of most organisms, contains several ghosts of ancient retroviruses, capable of jumping from place to place and carrying other genes with them. These "jumping genes" are called *transposons* or, more specifically, *retrotransposons*. Part of their role in evolution might lie in their capacity to reshuffle genes and gene control mechanisms. Another part might lie in their ability to make genes jump across species. For example, genes have been exchanged between humans and tsetse flies in Tanzania thanks to the activities of retrotransposons. It is curious to realise that retroviruses play a positive part in evolution. The most familiar example of a retrovirus is HIV, which has a bad press.

A retroviral gene (V) next to another gene (G) in chromosome A is transcribed to produce a messenger (mRNA):-

This messenger is then reverse-transcribed to make a new piece of DNA, which is then inserted into a different chromosome (B).

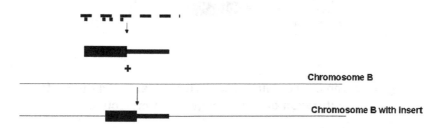

Fig. 11-1: how a retrotransposon moves from one chromosome to another.

Introns
DNA is stable and well protected, but it is nowhere near as passive and static as was believed before the 1980s. In this chapter we have surveyed several recently-discovered ways in which it exhibits plasticity. These discoveries have transformed the theory of evolution.

Another unexpected feature of genes was revealed during the late twentieth century: in eukaryotes, they are not usually *continuous* stretches of DNA. We mentioned this in a footnote to chapter 7. The protein-coding sequences (*exons*) are separated by pieces of DNA (*introns*) that have quite different functions, or are mere "junk". During transcription, the whole length of DNA encompassing the gene is "photocopied". The transcripts of the introns have to be cut out of the RNA after transcription. Then the exon transcripts are spliced together to make the messenger. For instance, consider a gene with a promoter (P), three exons (E_1, E_2 and E_3) and two introns (I_1 and I_2):-

Fig. 11.3 Intron splicing

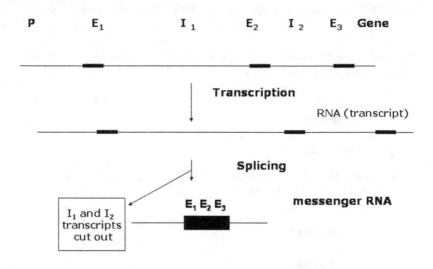

Fig. 11-2: removal of intron transcripts by splicing during the maturation of a messenger RNA molecule.

On the face of it, this seems a wasteful, cumbersome way of storing and expressing genetic information. Introns probably make up 80-90% of the total DNA in most eukaryotes, accounting for most of the "junk" content we alluded to earlier in this chapter. Why do introns exist? Why has evolution not eliminated them, streamlining the genome?

There are four plausible answers. First, some introns *do* contain useful information. For instance, they encode some small RNA molecules with exotic but essential functions in the cell. Second, an excess of non-coding DNA buffers genes against mutation, a point we mentioned earlier: the crowd protects the individual against the trigger-happy psychopath. Third, protein variants can be produced by *alternative splicing*. One of several alternative exons is used to make the messenger; the other alternatives are dumped. There are many examples of alternative splicing in humans. For instance, the receptors for a neurotransmitter might exist in several variant forms, but these are often encoded in a single gene. The gene transcript is alternatively spliced to give several different messenger variants. From these, the various receptor proteins are made. Thus, a single gene can encode several related proteins rather than only one.

The fourth possible advantage of the intron-exon arrangement in genes is evolutionary. Suppose one part of a gene, an exon, is transposed to a different place in the genome. (A cut through an intron does no detectable harm to the coding sequences.) The exon can then become part of another gene, generating a significantly modified protein that might have a useful new function. Evidence that such *exon shuffling* has actually happened in the evolutionary past is indirect, but certain domains in otherwise different proteins are remarkably similar; so it could be quite a common process and might have played a significant part in evolution.

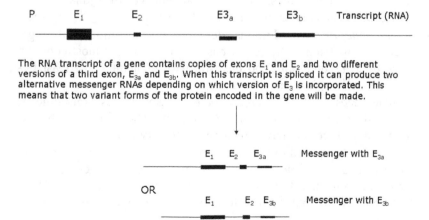

The RNA transcript of a gene contains copies of exons E_1 and E_2 and two different versions of a third exon, E_{3a} and E_{3b}. When this transcript is spliced it can produce two alternative messenger RNAs depending on which version of E_3 is incorporated. This means that two variant forms of the protein encoded in the gene will be made.

Fig. 11-3: alternative splicing; how two (or more) different proteins can be made from a single gene.

Genes and evolution

The mechanisms of DNA change that we have surveyed in this chapter help to explain how organisms have diverged so dramatically since the origin of life. The result is the incredible diversity of species in the modern world. Nevertheless, species that appear to be very different still have a remarkable amount of DNA in common. It is well known that we share more than 99% of our DNA with chimpanzees and scarcely less with gorillas, but it is a less familiar fact that we share 50% of our genes with bananas and 30% with baker's yeast.

Some of the genes that seem most clearly associated with multicellularity seem – surprisingly – to have originated in protists. Examples include hormone and hormone receptor genes. In large animals such as humans, a hormone is made in one specialised part of the body and dispatched to others, where it alters specific activities in target cells. What use are such devices in protists? What good are insulin and insulin receptors to amoebae? Whatever these gene products do in protists, it is obviously not the same thing that they do in mammals. During evolution, it seems, proteins can be recruited to different uses depending on the requirements of the organism. But to date, this fascinating topic has not been sufficiently researched to provide us with general answers.

Summary

Every cell protects the integrity of the genome. DNA is chemically stable, and the cell defends it with various protective and repair devices. When genes *are* altered, the consequences for the cell might not be disastrous: the system as a whole is robust, and the redundancy of the genetic code provides a buffer. So, in eukaryotes, does the abundance of "junk" DNA. Nevertheless, DNA can change in a number of ways.

- *Point mutations* are additions, deletions or substitutions of single bases, or inversions of pairs of bases. These are the most common kind of gene alteration, and the kind with (generally) the least far-reaching effects in terms of evolution.
- Whole genes can be *excised* and *deleted*. Sometimes excised genes are re-inserted back to front (*inverted*). These are much less common changes but have greater evolutionary impact.
- Genes can be moved to a different part of the genome (*transposed*), perhaps because of old retrovirus residues. They can even be moved from one organism to another, crossing species boundaries. These are rare events but their effects are cumulative and potentially dramatic.
- Numerous repeats of a single gene, or of a short DNA sequence, can be inserted into the genome. This is known as *amplification*. When a short, meaningless DNA sequence is reiterated the result is a

length of *simple-sequence DNA.* Simple-sequence DNA is "junk", but amplified genes can be essential for the cell's viability, for example ribosomal and histone genes.

* Most eukaryotic genes consist of one or more coding sequences (*exons*) separated by non-coding segments (*introns*). Exons might sometimes be moved from one gene to another to create novel genes and hence novel proteins. This process is known as *exon shuffling*.

These changes can alter organisms subtly or markedly. If a developmental "master gene" is modified, the effect can be particularly dramatic. These "master genes" encode transcription factors and control the expressions of many other genes, so significant changes in them can alter the entire course of development, perhaps creating a novel sort of organism.

Over thousands of millions of years, cumulative changes in DNA have resulted in the diversification of countless species. One result of the revolutionary progress in molecular biology during the last quarter of the twentieth century was a revision of the "tree of life". Before the 1970s, the historical connections among species were reconstructed mainly from comparative anatomical and embryological evidence and the fossil record. Now there is ever-increasing reliance on comparative gene sequencing. This modern, molecular biological, approach gives results that are reassuringly consistent with those of the traditional approach. However, comparative gene sequencing allows greater refinement of detail and has provided insights into early life on Earth, even into times before the fossil record began.

Chapter 12

THE SPICE OF LIFE
Diversity, natural selection and symbiosis

In chapters 2-10 we built up a general abstract model of "livingness", to which we believe all organisms and all cells conform. This model is about the *unity* of life. It describes what is common to all living things and distinguishes them from the non-living. But an adequate science of life needs to account for *diversity* as well as unity, and since 1859 this need has been met by the theory of evolution. The explanation of diversity is the primary role of evolutionary theory, though it is not the only role. Evolutionary ideas have been assimilated into all areas of biology, including molecular biology. But we encounter the theory directly when we seek to explain diversity and the occupation of the world's vast range of habitats.

The perspective we adopt in this chapter is quite different from earlier parts of the book. The focus now is on large collections of organisms, not cells and cell constituents. The theory of evolution tells us that organisms change by adapting to alterations in their environment, yet they are still organisms. The theory is itself ever-changing, adapting to advances in knowledge, but at root it is still the same theory.

To paraphrase John Donne, no organism is an island. Organisms exist in breeding *populations,* not as isolated individuals. Moreover, populations of different species interact: they eat one another, inhabit one another, transport one another, depend on each other's waste products, spread diseases to one another or simply compete for space. In a given geographical area, these interactions constitute an *ecosystem*. Ecosystems differ, but each comprises a more or less wide diversity of organisms, sometimes hundreds of thousands of species. The individual belongs to a population that is part of an ecosystem, which in turn is part of the *biosphere*, the part of the planet that houses life.

Diversity

Attempts to classify plants and animals (*taxonomy*) date back to Classical Greece. Modern taxonomy was pioneered by Linnaeus in the 18th century. Linnaeus listed several thousand species of plants and animals and introduced the now-standard system of nomenclature: a double-barrelled Latin name comprising the genus (type) and species, as in *Homo* (genus) *sapiens* (species). Genera were grouped into families, families into classes, classes into orders, and the orders into one of two kingdoms, plant and animal. Animals were divided into vertebrates (mammals, birds, reptiles, amphibians and fish) and invertebrates (all the rest).

A number of difficulties have become apparent in Linnaeus's system. During the 250 years since it was published the number of known species has grown to millions. The total number of animal species *believed* to exist is estimated at thirty million. Then there are all the plants, fungi and single-celled eukaryotes; and the prokaryotes, the most ancient and most numerous forms of life on Earth[23]. When we consider *all* organisms – animals, plants, fungi, protists and prokaryotes – the number of distinct types in the world today might be around a hundred million. (We use "types" here rather than "species" because of the difficulty of applying the word "species" to prokaryotes; see footnote 23.) That is the world *today*. It is generally presumed that at least 99% of all species that have ever existed are now extinct; extinction seems to be the ultimate fate of all. This suggests that the number of types of organism that have existed since the origin of life may be around ten thousand million (10^{10}). The figure might be inaccurate but the implication is clear: the variety of ways of being a viable organism is staggering.

Because we know so much more about species, past and present, than was known to Linnaeus, taxonomy has changed dramatically through the intervening years. Late in the 19th century, the German evolutionist Haeckel suggested a third kingdom, monera, distinct from animals or plants and comprising all microscopic organisms. Haeckel also added another level of

---

[23] During the 1970s, Erwin identified 1100 species of beetle inhabiting one species of Amazonian tree and he estimated that 160 of these beetle species were unique to that tree. Extrapolating from these figures, 50,000 species of tropical trees implies 50,000 x 160 = 8 million species of beetles. Given that beetles seem to account for a quarter of *all* animal species, this suggests that there may be about 30 million species of animals. As for prokaryotes, it is debatable whether we can use the term "species" for them as we can for eukaryotes. They can exchange genes, singly or in small groups, almost at liberty, which makes the concept of a species-related genome inapplicable. However, it is generally assumed that there is more divergence among prokaryotes than there is among eukaryotes as a whole - which would not be surprising, given their relative venerability as tenants of the Earth.

classification, the *phylum*, intermediate between order and kingdom. Humans are now classified as: kingdom – animal; phylum – chordates (vertebrates); order – mammals; class – primates; family – hominids; genus – *Homo*; species – *sapiens*. Early in the 20th century, Copeland divided Haeckel's monera into protists (single-celled eukaryotes) and prokaryotes. In the 1950s, Whittaker suggested that, because fungi are fundamentally different from plants, they should be regarded as a separate fifth kingdom. Most biologists today consider every living organism to belong to one of these five kingdoms.

In the 1980s, Woese divided prokaryotes into (a) what are now called *archaea* and (b) true bacteria. This distinction is based on a single ribosomal gene, but Woese considers it so dramatic, so fundamental, that archaea and bacteria differ as much from one another as they do from eukaryotes. Not many biologists entirely agree with Woese, but the distinction between archaea and bacteria is generally recognised. These organisms look similar but biochemically they are very different. Nevertheless archaea can exchange genes with bacteria, just as bacteria do with other bacteria.

One effect of the taxonomic revolution has been to marginalise our own species. Once upon a time there were two kingdoms, one of which (animals) was divided into vertebrates and invertebrates. We were a significant example of the vertebrates. This was a hopelessly biased classification. It ignored microscopic life, which is most of life. Moreover, dividing the animal kingdom into vertebrates and invertebrates was like dividing the world's land masses into the Isle of Malta and Not the Isle of Malta. Now there are five kingdoms, of which the animal kingdom by no means the biggest and is divided not into two sub-kingdoms but some thirty-five phyla, of which chordates (vertebrates) are just one. Alternatively, if Woese's taxonomy is accepted, there are three super-kingdoms, and animals *in toto* make up one subdivision of just one of these.

The range of habitats
Why are there so many different ways of being alive? Why is there such a huge, apparently limitless variety of organism types, all (so far as we can tell) conforming to the general abstract model of the living state that we advanced in chapter 10? One answer is that organisms inhabit a wide variety of environments. Over the aeons, changes in DNA have led to the emergence of countless different types of organisms, differing enormously in character. Collectively, these organisms have come to inhabit virtually every kind of environment that the Earth can provide.

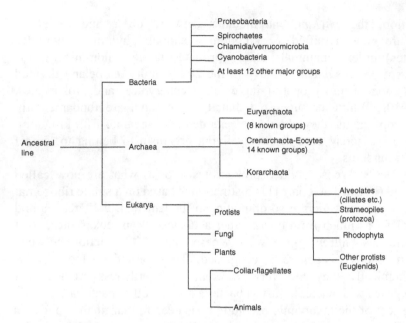

**Fig. 12-1: an outline scheme of the phylogenetic tree based on
Woese's 'three kingdoms' model.**

Life is ubiquitous on and near the Earth's surface. Every conceivable
niche is exploited by organisms with appropriate lifestyles. Under the
Antarctic ice cap, in the middle of the hottest and driest deserts, in volcanic
vents in the depths of the oceans, even in the clouds, there are organisms.
The polar seas swarm with archaea; bacteria live and reproduce in cumulus
clouds, which have life-times in the order of several days[24]. Prokaryotes live
in hot rocks deep in the Earth's crust. Each organism, no matter where it is
found, is equipped to survive in its particular environment. Each is part of a
local ecosystem.

---

[24] The discovery that there are about 1500 bacteria per millilitre of cloud meltwater is recent.
The bacterial types have not, at the time of writing, been identified. It has been known for
some time that bacteria can be transmitted from place to place on the Earth's surface via
the upper atmosphere, but the discovery that bacteria actually *live* in clouds - at sub-zero
temperatures in high-intensity ultraviolet radiation - is remarkable. What do they live on?
Organic matter thrown up into the clouds in sea-born droplets is one possibility. And do
they affect the weather, e.g. by affecting cloud formation and rainfall? Do they account
for the fact that clouds are more opaque to sunlight than their water contents entitle them
to be? The questions attendant on this - as on any striking new discovery - are legion.

   Most television wildlife programmes make the point that the world teems with life. But television programmes often concentrate on macroscopic rather than microscopic life, the minority not the majority, so they do not tell the viewers how extreme some environments are. No plant or animal can survive for long at temperatures of more than 50°C and no protist at temperatures of more than 60°C, or at pressures greater than a very few atmospheres. But certain archaea live in volcanic vents at pressures of hundreds of atmospheres and temperatures well in excess of 100°C (for example, *Pyrolobus fumarii* lives at 113°C), and many of them enjoy conditions so acidic that most other organisms would die within seconds. There are numerous examples of apparently lethal environments in which life flourishes. *Thiobacillus ferrooxidans* thrives on the iron in reinforced concrete; it breeds furiously in modern architectural masterpieces and makes bridges and tunnels collapse.

Fig. 12-2: a hydrothermal vent ("dark smoker").

The theory of evolution: an outline

Briefly, the theory of evolution in its classical form is based on a few near-commonplaces. (This "classical form" was fully articulated in the 1930s.)

1. In any population of a species there is variety - no two individuals are alike (even identical twins are exactly alike only in their genomes).

2. At least some of this variation is genetic. In chapter 11 we surveyed the ways in which genetic variation can be generated.

3. Variation means that some individuals in the population are better able to survive and leave offspring than others, though the differences might be very slight.

4. As a result, the variant genes that favour survival become relatively more abundant in the offspring generation. This is the process known as "natural selection".

5. Over a series of generations, therefore, the genes for more successful variants become progressively more abundant in the population's gene pool[25].

6. However, the individual's survival chances depend on an unlimited number of environmental factors (climate, food resources, predation, etc.). Therefore, whether a particular variant of a gene favours survival and reproduction is a matter of time and place. It depends on the nature of the ecosystem.

7. In other words, gene variants that are good for survival in one ecosystem might not be good in another. Therefore, populations of the same species in two different ecosystems diverge genetically over time. And the gene pool of a population within one geographical area changes as environmental conditions change.

Viable scientific theories change with time because they adapt to new discoveries - much as viable biological species change over generations because they adapt to changing environments[26]. Since its inception in the 19th century, the theory of evolution by natural selection has undergone serial modifications. Its most radical adaptations have been the assimilation of classical genetics and, more recently, molecular biology. It continues to develop amid healthy and sometimes heated scientific debate. We shall say more about this aspect of the history of science elsewhere. Here, we concentrate on the theory itself.

---

[25] "Gene pool" is to population as "genome" is to individual. It means the sum total of all the genes (each in all its variant forms) in all living individuals in the population.

[26] More than one philosopher of science has explored this analogy between the evolution of scientific theories and the evolution of species. Stephen Toulmin has made particular use of it.

Diversity is important for life

To explain the incredible diversity of life is one thing; to understand why diversity is important for life is another. What difference would it make if the world only contained a few thousand species rather than untold millions? In the *Origin of Species* Darwin observed:

> If a plot of ground be sown with one species of grass, and a similar plot be sown with several distinct genera of grasses, a greater number of plants and dry herbage can be raised in the latter than the former case... The greatest amount of life can be supported by the greatest diversification of life.

Generations of research since the publication of this classic have supported Darwin's assertion. During the 1950s, Elton reviewed a number of ecological studies of many different habitat types. He concluded that the greater the diversity of organisms within an ecosystem, the more stable and productive the ecosystem is. His pronouncement echoed Darwin's. More recent experimental work points to the same conclusion.

This adds something to the idea of natural selection (competition *within* species), which was summarised in the previous section. It adds the idea of co-operation *among* species. This use of the word "co-operation" does not imply a conscious act or altruism. It means that the presence of some species helps other kinds of species to become established and flourish. As we remarked at the start of this chapter, organisms need other species because they eat one another, inhabit one another, transport one another, depend on each other's waste products and so on. The continuing survival of any species depends on the continuing survival of other species. It is a simple step to the conclusion that ecosystem diversity creates stability.

A cell is made of huge numbers of molecular components linked in a network of mutual dependence. We explored this in chapters 2-10. The network of mutual dependence confers stable order on the cell. Similarly, an ecosystem is stable and ordered because the species it comprises are linked by a network of interdependence. The more components there are, and - up to a point - the more interdependence there is among them, the more ordered and stable the network. This is a principle of complexity theory. According to one authority on complexity theory, Stuart Kauffman, both cells and ecosystems are ordered networks - but not *too* ordered. If a large network is too tightly interconnected, its behaviour becomes fixed and "frozen". On the other hand, if its interconnectedness is insufficient, it behaves chaotically. Kauffman argues that evolution drives biological networks towards "the edge of chaos", a state that is ordered but borders on the chaotic. In this state, behaviour is flexible and adaptable, stable and ordered but not rigid. His argument is persuasive.

So far as the interconnectedness of ecosystems is concerned, the *intimacy* of some relationships among species is surprising. Some flowering plants are pollinated by particular species of flying beetles. Without the beetles the plants would die out; they could not reproduce. Many of these plants, such as the tropical lily *Philodendron selloum*, maintain their flowers at 35°C, an ideal temperature for keeping the flight muscles of the resident beetles working. If it were not for the warm flowers, some types of beetles (which are very small animals) would have to eat their own weight of food every day in order to generate enough heat to fly - an impossible demand. So the beetles are as dependent on the flowers they pollinate as the flowers are on the beetles. If the flowers die out, so do the beetles - and vice-versa. Many comparable examples of intimate plant-insect relationships have been described, most of them species-specific.

Many species interrelationships are even more intimate. Organisms of different types are often locked in an absolute mutual dependence known as *symbiosis*. Lichens are familiar examples. A lichen is a symbiont comprising a fungus and a green alga; neither species can survive independently of the other. Lichens are very widespread; it is estimated that the world contains about 10^{14} tons of them. Herbivorous mammals such as cows and rabbits, and wood-devouring insects such as termites, eat vast quantities of cellulose, which they cannot digest; but bacteria in their guts digest the cellulose and thus provide themselves and their hosts with nutrient. Without these bacteria, most herbivores would starve; and if herbivores starved, so would the rest of us. Without termites, ant-eaters would go hungry, though the wooden structures we build in the tropics might be more durable. Without the herbivores (or termites) to provide comfortable guts to inhabit, the cellulose-digesting bacteria would not survive either. A cow is a symbiont. A termite is a symbiont. The average tree is estimated to contain some 300 fungal symbionts. The first plant is believed to have resulted from symbiosis between a fungus and an alga some 430 million years ago.

The more closely we examine the huge variety of plants, animals and fungi in the world, the more examples of symbiosis we find. Symbiosis seems to be the rule in life, not the exception. Therefore, natural selection *within* species takes place in a world of mutual dependence - usually intimate, often symbiotic - *among* species. **Life progresses through competition and extinction and survives through co-operation and diversity**.

(Some organisms, *parasites*, come to depend on others without providing anything in return except discomfort and possibly ill-health. Co-operation might be the rule in all ecosystems, but some organisms exploit rather than co-operate.)

Summarising the theory of evolution in the previous section, we described the crucial role of "environmental factors". The environment selects among variants of a species, enabling some to survive and reproduce more successfully than others. What *is* the environment? Clearly, it is not just the physical environment; it must include the rest of the ecosystem. And what exactly is selected? The individual? But in many cases the "individual" is a symbiont: two or more species, not one. In any case, the survival of a variant of one species has implications for the rest of the ecosystem. If one element of the network changes, the others must also change; the network adapts. The network of mutual dependence survives, but it is not the *same* network from generation to generation. Moreover, groups that co-operate are favoured by selection. "Cheats" that welsh on their part in co-operation and take advantage of the rest do not predominate. So although many parasites are successful organisms and survive, "cheating genes" do not generally seem to be favoured[27].

Should the population be regarded - or the ecosystem as a whole - as a unit of selection? Populations change (diverge) by natural selection among individuals; species change because their populations change; ecosystems change because species co-evolve. On the other hand, some writers have argued that *genes* are the units of selection; the environment of a gene is other genes. We would not go along with this last point of view, but it has a respectable following. The fundamental concepts here, "selection" and "environment", are difficult to pin down. More than one meaning can be ascribed to each.

The Gaia hypothesis
We introduced the Gaia hypothesis briefly in chapter 6. According to this hypothesis, the physical environment alters life via natural selection, but life also alters (some would say "regulates") the physical environment. The original proponent of the idea, James Lovelock, observed that the

---

[27] "Group selection theory", advanced by Trivers in 1971, has always been controversial and some writers, not least Richard Dawkins, are strongly antipathetic to it. Dawkins says that everything attributed to group selection can be accounted for by kin selection (behaving altruistically towards close relatives, which share a large percentage of your genome); kin selection theory was pioneered by Hamilton in 1964. Nevertheless some recent ecological evidence, surveyed for example by Sober and Wilson in 1998, is difficult to explain without recourse to group selection.

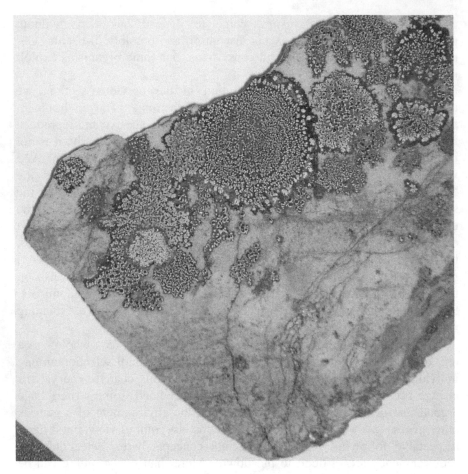

Fig. 12-3: a lichen, a symbiont comprising an alga and a fungus.

atmospheres of Mars and Venus are very close to chemical equilibrium, but the Earth's atmosphere is very far from equilibrium. The difference arises, according to Lovelock, because there is no life on Mars or Venus to alter the composition of the atmosphere. The composition of the Earth's atmosphere remains stable for very long periods, though its component molecules are rapidly exchanged with the planet's surface. The oxygen content is maintained at a level sufficient for the needs of large oxygen-dependent organisms such as ourselves, but not high enough to cause widespread fires, which would seriously disrupt life. *Life* keeps the atmosphere stable, yet far from chemical equilibrium.

Another argument in favour of Gaia is that purely geochemical explanations of the Earth's climate history seem inadequate. For example, the reflection of sunlight from the ice caps should lead to intermittent total glaciation of the planet. This seems to have happened only very occasionally, if at all. The near-constancy of the Earth's average surface temperature over extended periods is remarkable; the sun's power has increased by some 25% over the past two thousand million years. The only influence that can have buffered the planet against the expected gradual heating and intermittent global cooling, say the proponents of Gaia, is the presence of life.

Calcium liberated from silicate rocks by weakly acid rain reacts with carbon dioxide to form calcium carbonate (limestone or chalk). This process removes carbon dioxide from the atmosphere, making the rain less acidic. This in turn decreases the weathering of silicate rocks. These days, everyone knows that carbon dioxide is a greenhouse gas; as its levels in the atmosphere rise and fall, so does the temperature near the surface of the planet. So the rate of silicate rock weathering and limestone formation are linked to the surface temperature via atmospheric carbon dioxide levels. However, the processes just described are much too slow to account for the geological evidence. The involvement of organisms in removing atmospheric carbon dioxide by photosynthesis and regenerating it by respiration and decomposition accounts more credibly for the speed of geological events.

Contrary to what some critics have suggested, the Gaia hypothesis is consistent with the natural selection model of evolution. In fact, it extends the argument we discussed in the previous section (different species within an ecosystem influence one another). The growth rate of any population depends on environmental variables; life defines the tolerable limits of such variables. Natural selection determines which species dominate the environment at any given time. By definition, these dominant species exert the greatest effects on the environment. So the environment determines which species predominate, and the dominant species alter the environment. Organisms and environment are tightly coupled and co-evolve. If a "pro-Gaian" and "anti-Gaian" mutation were to arise simultaneously within a population, then the "pro-Gaian" ones would be preferentially selected. Computer simulations support this prediction.

Beyond doubt, the Gaia hypothesis is valid to this extent: life affects the environment just as the environment affects life. This principle helps to explain the Earth's geological and climatic history and leads to predictions that can be tested by simulation.

However, it is unwise to extrapolate the hypothesis too far. The influence of life might not *always* accelerate or retard geological ageing to a significant extent. Also, it is not clear that the whole Earth behaves homeostatically, as some proponents of Gaia would claim. There are no plausible grounds for considering the whole planet to be a unit of selection or a "superorganism", as some enthusiasts have proposed. The Gaia hypothesis is valuable, but we should not transmute it into mysticism.

How much of a lottery is evolution?
How much of the history of life is a matter of chance? In some ways the answer must be "a lot". Alterations of DNA such as point mutations are random, so the generation of new variants in a population is a random process. Natural selection is emphatically *not* random - it is a very precise process, capable of favouring one variant over an infinitesimally different one - but it depends on environmental factors that are largely unpredictable. All these points are generally agreed among biologists.

However, opinions differ about the range of *possible* forms of life. Gould argues that the history of life is almost entirely contingent. Rewind the tape of evolutionary history and play it again as many times as you will; it will never be the same twice. If life on Earth began and evolved all over again, there would be no dinosaurs, no fish, possibly no eukaryotes, and certainly no humans or bananas. This is because DNA changes entirely at random, so body plans can change in a virtually unlimited number of ways at every evolutionary step. The overall process is completely unpredictable.

In contrast, Kauffman argues for a finite number of possible body plans; a finite number of solutions to the problem of being alive and reproducing. If the tape of history were rewound and played back, then it could look similar - though not identical - on each re-run. This is because an adaptive complex system, such as life evolving on Earth, produces order as it goes along. Kauffman's position here is consistent with the Gaia hypothesis. He reasons that the "phase space of evolution" (the set of all possible as well as actual living forms, past and present and future) is not fixed; it evolves in response to the actual organisms it encompasses. The self-organising nature of life at the organism/ecosystem level cannot be seen from the molecular biological standpoint, but it is undeniable: it is a mathematical property of complex systems. In practical terms, the "Kauffman mechanism" might work through symbiosis (causing "jumps" to higher levels of organisation), and through co-evolution with a continuously changing environment. According to Kauffman, it results in a much more limited range of possible body plans than Gould would allow, though what is "possible" changes over time.

In Gould's favour one can cite the succession of extraordinary organisms that once lived on the Earth but have vanished with all their kind; no subsequent species has ever resembled them. Evolutionary novelty in body plans is unpredictable. One can also cite the astonishing diversity of species extant today. In Kauffman's favour is the indisputable fact that organisms have become more and more complicated as time has gone on; hardly a random process, but probably the result of successive symbioses. There are also many cases of convergent evolution, such as the close anatomical similarity between mammals that have returned to the sea (whales, for example) and fish.

Biologists argue for a range of positions between the "extremes" represented by Gould and Kauffman. This is as it should be. Too much agreement is a bad thing; debate is the life-blood of science. No serious biologist today doubts the theory of evolution, but the details of how evolution works - what constitutes the environment, what the unit of selection is, how contingent the process might be – remain the subjects of healthy and productive argument, and promise to remain so.

Chapter 13

CURRICULUM VITAE
An outline history of life on Earth

Changes in DNA result in variations among offspring (chapter 11). As a result, variant organisms achieve greater or lesser reproductive success depending on interactions with their local population, with partners in symbiosis, and with their environments (chapter 12). These processes have produced the vast diversity of organisms from a common ancestor. It has taken an immense amount of time for them to do so.

Fossil and radioisotope dating evidence along with comparative DNA sequencing tell us that life on Earth has had a continuous history of at least 3,800 million years, some 80-85% of the age of the planet. For the first two thirds of that time, all organisms were prokaryotes. For 70% of the time there were no multicellular organisms. Roughly 85% of that vast period had elapsed before any recognisable animals appeared, 90% had passed before there were land plants, and 99.9% before the earliest hominids. Future discoveries might alter these estimates but they are approximately correct. However, numbers cannot convey a mental picture of the huge time-scales involved. The following diagrams might be more helpful; the scales are thousands of millions of years before (minus sign) or after (plus sign) the present.

Our sun is a second-generation star. That is, the sun and its planets, including the Earth, were made from debris from the disintegration of first-generation stars, which were formed after the Big Bang. The chemical elements necessary for life as we know it, such as carbon and oxygen, were made as the first-generation stars neared the ends of their lives. The sun was formed about 5,000 million years ago and the planets during the succeeding 500 million years. A further 500 million years probably elapsed before the Earth could support life, which it has continued to do ever since. It seems likely that the Earth will go on supporting life for another 1,500-2,000 million years. A star such as the sun survives for about 10,000 million years altogether, so about half its life-span has elapsed so far.

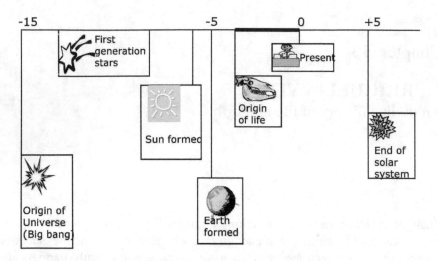

Fig. 13-1: overview of the lifespan of the solar system, and of life on Earth, in relation to the age of the universe.

These facts are relevant to the origin of life on Earth and elsewhere, matters to which we shall turn in chapters 14 and 15. But for the present chapter we need a scaled-up version of the highlighted segment of Fig. 13-1: the interval between the origin of life on Earth and the present day is shown in Fig. 13-2.

What have been the most salient events in this 3,800 million year history? What we consider "salient" is largely a matter of viewpoint, and our viewpoint is unavoidably anthropocentric. It also depends on the evidence available to us. The right-hand part of Fig. 13-2 (nearest the present day) is relatively crowded. This might be because more has happened recently, as organisms have grown more diverse and complex. But there is a simpler reason: we know a lot more about recent times than ancient ones. The diagram is unbalanced not (necessarily) because very little happened during the first 70% or so of life's history, but because we are largely ignorant of it.

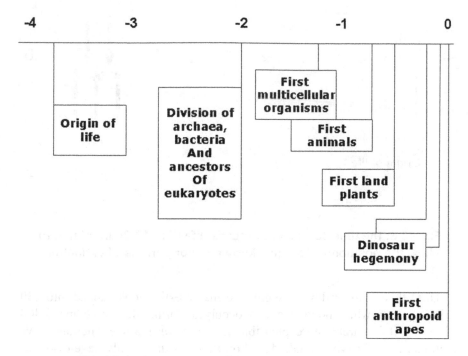

Fig. 13-2: time course of the history of life on Earth.

To illustrate this, consider our classification of rocks on the Earth's surface into geological eras: Cambrian, Ordovician, Devonian and so on. These eras segment the last 600-700 million years, the segments becoming smaller as they approach the present. Everything older than that – in other words, 85% of the history of the planet – is labelled "Precambrian". We can only base our classifications on evidence, including evidence of what was living when the rocks were formed, and we have very little basis for subdividing the "Precambrian".

The way we describe periods of mass extinction also shows how selective our knowledge is. There is compelling fossil evidence that five major periods of extinction have punctuated the history of life. These were intervals of little more than a million years during which large percentages of existing species were wiped out. In the third and biggest of the five, which occurred at the end of the Permian era, 96% of all known species disappeared. In the wake of each of these catastrophes, the ecological vacuum was filled by newly-evolved species and the Earth was re-populated with novel organisms. If we locate these five major extinctions on our second time diagram, we obtain the following picture:-

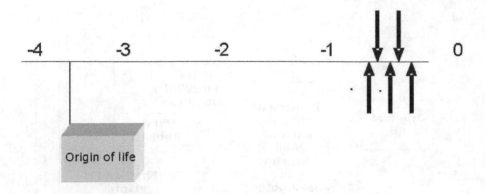

Fig. 13-3: the time course of terrestrial life (Fig. 13-2) annotated to show the positions of the five known major periods of extinction.

Does this seem credible? Were five major extinctions packed into 430 million years, while no extinctions occurred during the preceding 3,400 million? It is surely more plausible that *regular* mass extinctions have occurred, roughly every hundred million years. But we only have evidence for the last five, not the previous thirty or so.

The conclusion is inescapable. Our attempts to reconstruct the first three-quarters of the history of life on Earth are based on very scanty knowledge, and are probably wildly inaccurate.

The earliest organisms
What was the Earth like when life first appeared? It was probably very hot; there would have been liquid water, but it was probably close to boiling point over most of the surface. The planet's crust was thin; volcanoes and earthquakes were commonplace. Meteorite storms and comet collisions were regular events. Thunderstorms were probably ubiquitous and virtually continuous. The atmosphere was mainly carbon dioxide and nitrogen, mixed with gases emanating from volcanoes. There would have been virtually no free oxygen. Because there was no free oxygen there was no ozone layer, so if and when the cloud cover was not too thick, ultraviolet radiation penetrated freely to the surface.

We can only guess about the Earth's first inhabitants. They were presumably small prokaryotes with the simplest possible genomes. They probably resembled archaea rather than bacteria because they had to survive in a very hot and acidic environment. Oxygen would have killed them; oxygen is very damaging to living materials. A large percentage of present-day organisms tolerate it, even require it, only because their cells have elaborate mechanisms for rendering it harmless. When life first appeared on the planet there might have been some free hydrogen in the atmosphere. (Hydrogen is quickly lost from the atmospheres of small planets because it is such a light gas; strong gravitational fields such as those of Jupiter and Saturn are needed to retain it.) If so, the first organisms might have used hydrogen as a fuel source. Otherwise they probably used hydrogen sulphide released from volcanic vents, or similar chemical reductants. How many "species" of these earliest organisms there were, how many arose independently from non-living sources, we cannot guess. But since comparative DNA sequencing evidence points to a single common ancestor, all but one of these "species" must have vanished as though they had never been.

Because the earliest organisms were probably archaea-like, some people conjecture that the prokaryotes that inhabit deep ocean vents today must resemble them. These deep ocean vent organisms are therefore living fossils of the very earliest life. As we mentioned in chapter 12, they tolerate high temperatures, high pressures and high acidity and utilise hydrogen sulphide. The possibility is reasonable, but it would be a mistake to *equate* modern archaea with the pioneers of terrestrial life. Modern archaea seem to require products of photosynthesis, which reach them from the ocean surface. Geological evidence shows there was no oxygen in the atmosphere when life began. Photosynthesis produces oxygen, so if there was no oxygen in the atmosphere, there was no photosynthesis. Therefore, modern archaea are not identical with the earliest organisms. Atmospheric oxygen became detectable later, and the amount increased slowly over the succeeding two thousand million years.

The earliest organisms probably obtained their energy from chemical sources such as hydrogen sulphide, but could they have used sunlight? Some modern archaea and bacteria utilise solar energy not by photosynthesis, but by a simpler mechanism that does not produce oxygen. Sunlight activates a membrane protein, which pumps hydrogen ions out of the cell. The resulting hydrogen ion gradient is used to manufacture ATP, the "common energy currency" of metabolism (chapter 4). This simple light-driven hydrogen ion pump operates in many present-day archaea and bacteria, so it might have a very ancient origin. The mitochondria and chloroplasts of eukaryotic cells, which were once free-living prokaryotes, also use hydrogen ion pumps to couple their energy sources to ATP synthesis.

Photosynthesis

Before many millions of years had passed, some cells had learned to use sunlight to split water. Rocks formed over 3,000 million years ago show that the atmosphere contained traces of oxygen by then. Photosynthesis had begun.

Photosynthesis opened an evolutionary door. Without an oxygen-rich atmosphere there could have been no animals. Most of the prokaryotes, protists and fungi alive today could never have evolved without photosynthesis; they too depend on oxygen. But oxygen is lethal to all living matter unless protective devices are installed. Every oxygen-breathing or oxygen-tolerant species in the world is equipped with molecules designed to destroy deadly oxygen derivatives. The first organisms ever to develop photosynthesis, presumably primitive cyanobacteria, must have been able to detoxify the oxygen they generated. Otherwise, photosynthesis would have been suicide.

This simple deduction raises further questions. How did the earliest organisms detoxify oxygen? Perhaps they used a simple combination of amino acids, such as those associated nowadays with marine bioluminescence[28]. Alternatively, a copper or iron containing enzyme might have been used, as in most present-day terrestrial organisms. Whatever the mechanism, it must have evolved before photosynthesis, or at latest simultaneously with it; so why did it evolve before there was any apparent need for it? Perhaps, to begin with, the mechanism served some entirely different purpose. Or perhaps it did not pre-exist, but arose as a by-product when the prototype chlorophyll was synthesised. Could a single anabolic process have given rise to both a chlorophyll-like and a coelenterazine-like[28] molecule, creating photosynthesis and oxygen-protection in one fell swoop? The questions remain.

Did *all* early-Earth organisms develop the equipment to detoxify oxygen? Three scenarios can be imagined.

1. Most organisms alive at the time could detoxify oxygen; photo-synthesis emerged in a pre-prepared world. But if so, why? This takes us

---

[28] Amino acids are the building blocks of proteins. Amino acid is to protein as a single popper bead is to a necklace. A dietary requirement for most bioluminescent organisms today is a molecule called coelenterazine, which is made from three amino acids joined together. Coelenterazine penetrates into cells and reacts very quickly with oxygen, emitting a brief flash of light when it does so. Perhaps the original function of coelenterazine was to protect against oxygen toxicity. It is an essential part of the diet of many marine organisms today, not only luminescent ones (though bioluminescence is very widespread in marine life), but its source remains mysterious. No one knows what species makes it, or why.

back to the previous questions. Why would an oxygen-detoxifying mechanism have evolved in a world without oxygen? What function could it have served before there was any anything to detoxify? And what was it?

2. There were no other organisms. There was only one "species" (the photosynthetic one), or a very few closely related types. Therefore, life was presumably very localised and thinly spread. The early Earth ecosystem was seriously lacking in diversity, which means it was unstable. Therefore life had only a tenuous hold on the planet and was always likely to die out. Since life obviously survived, this scenario seems unlikely.

3. All other organisms were wiped out by the deadly new pollutant, oxygen, except for a few that survived in oxygen-free enclaves just as anaerobic organisms do today. This is the likeliest scenario. Photosynthesis unleashed a mass extinction 3,000 million years before the earliest one known from the fossil record. Perhaps this is why all extant organisms seem to have derived from a single common ancestor. There might have been several different origins of life, but only one lineage survived after oxygen appeared in the atmosphere.

Are we limited to these three scenarios? The atmosphere's oxygen content climbed only very slowly from its initial zero value. It took the better part of 2,000 million years to reach its present level. When photosynthesis first began, the resulting oxygen level was so minute that even a rudimentary protection mechanism would have sufficed. So protection against oxygen poisoning might have evolved slowly and gradually. Nevertheless, photosynthesis must have forced a separation between oxygen-tolerant organisms, later to become bacteria and eukaryotes, and oxygen-intolerant ones, later to become archaea. Yet Woese and his colleagues date the separation of the archaea from the bacteria and "eukarya" to *two* thousand million years ago, not three. Why did atmospheric oxygen take a thousand million years to force this separation? No one has the answer.

The first eukaryote

According to Lynn Margulis, symbiosis is fundamental to life. She explains the evolution of eukaryotic cells, multicellularity and the evolution of more complex organisms in terms of symbiosis.

Margulis's account carries conviction. Nothing about evolution is pre-planned. It follows no strategy, it has no goal, and there is no mysterious agency *driving* it towards ever-higher complexity. Every event is the consequence of antecedent causes, like everything else in the natural world. Evolution happens because of random DNA changes and the different reproductive successes of the resulting variant organisms. But the environment that determines success includes other organisms. The more intimate the interactions with these other organisms, the more profoundly they affect survival and reproductive success. So if symbiosis - the most intimate of interactions – is almost ubiquitous, its effect on the course of evolution must have been great.

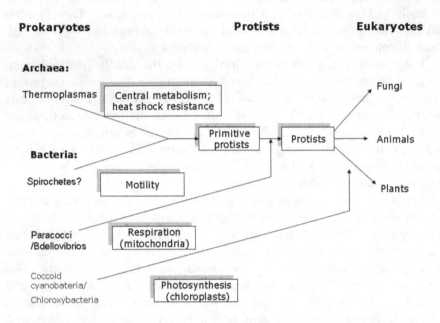

Fig. 13-4: a possible scheme for the evolution of eukaryotes, based on the argument proposed by Lynn Margulis.

Bacterial mats populated the shallower parts of the sea between one and two thousand million years ago. The constituent organisms were of various types ("species"), bound together in miniature ecosystems. For the most part they remained separate organisms, though no doubt they sometimes exchanged genes, as modern prokaryotes do; but these mats provided an ideal setting for symbiotic union. According to Margulis, a type of swimming bacterium known as a spirochaete[29] invaded one or more archaea, and the two different DNAs survived together in a combined cell. This is perhaps how the prototype eukaryote was formed.

Not everyone agrees with this scenario. Taylor and Cavalier-Smith, for example, believe that eukaryotes began through over-replication or

---

[29] Spirochaetes are long thin bacteria that are highly mobile. The best known example is the organism that causes syphilis, but this should not give spirochaetes a bad name. Most of them are entirely harmless to humans.

branching development of an archaeal DNA, and Hartman suggests that the eukaryotic nucleus was originally a symbiont that took up residence inside the ancestral protist. However, some lines of evidence seem to favour the Margulis model: (a) spirochaetes *do* invade archaea, usually destroying them; (b) biochemically, eukaryotes have roughly equal numbers of bacteria-like and archaea-like characteristics. Also, two different cytoskeleton proteins that predominate in all eukaryotes (see chapter 3) are remarkably similar in their properties but are the products of wholly unrelated genes (a molecular example of "convergent evolution"). This suggests that the earliest eukaryotic cell had two disparate forebears[30].

So there are different opinions about the origin of eukaryotes. However, there is almost universal agreement about what followed: a symbiotic fusion between the ancestral eukaryote and a protobacterium. The pioneering eukaryote could not have survived without oxygen metabolism. If it could not metabolise oxygen itself, it must have co-operated very closely with a bacterium that could. Many modern-day bacteria make ATP by transferring the hydrogen atoms obtained from nutrient molecules on to oxygen, producing water (chapter 4). If such a protobacterium gained nutrients from the proto-eukaryote and supplied ATP in return, that would have led to close co-operation. But the co-operation would not have been fully efficient until the protobacterium took up residence *inside* the pioneering eukaryotic cell, the most intimate symbiosis imaginable. This is how mitochondria are believed to have originated, guaranteeing copious ATP supplies for the cell.

Over the ages that followed, the "assimilated" bacterium lost most of its genome; nearly all its proteins became encoded in the host's nuclear DNA. But the loss of mitochondrial DNA has not been total. Even today, mitochondria have small circular DNA molecules of their own: typically prokaryotic DNA, but much smaller than the genome of any independently living bacterium. And to some extent they can replicate independently of the rest of the cell.

---

[30] One of the main types of cytoskeletal fibre (microfilaments, which are involved in cell movement and muscle contraction) is made of a protein called *actin*. Another major type (microtubules, which are involved in axonal transport and cell division) is made of a protein called *tubulin*, which has very similar properties to actin but is totally different in composition. Every known eukaryote contains both these proteins, which are encoded in wholly unrelated genes, suggesting that two unrelated ancestors were involved. Tubulin is the sort of protein that *might* be found in a spirochaete because it is central to many of the swimming devices found in many cells (cilia and flagella), though no known spirochaete actually does contain it. Nor does any known archaeal species contain actin. But this is not compelling evidence against the hypothesis. Margulis could be right: the ancestral species might have died out during the past thousand million years or so, in which case further evidence for the origins of these proteins is unlikely to be forthcoming.

The protobacteria that gave rise to mitochondria were probably the ancestors of many modern-day free-living bacteria. They also gave rise to Rickettsia, intracellular parasites that in some ways resemble mitochondria. Rickettsia cause human diseases such as typhus. Their intracellular parasitism shows that organisms of this type *could* have taken up residence in an ancestral eukaryote. In some early eukaryotes, a similar process probably resulted in the uptake of symbiotic cyanobacteria into the cell. These were the first chloroplasts. Chloroplasts, like mitochondria, have small residual circular DNAs of their own even today. And free-living cyanobacteria abound in the modern world.

We can now see why there are no "half-way houses" between eukaryotes and prokaryotes. We raised this question in chapter 3. No matter how the first eukaryotic nucleus was formed, everyone agrees that it must have contained at least twice as much DNA - twice as many genes - as a prokaryote. Moreover, it assimilated the ancestors of mitochondria, leading to still greater metabolic and structural elaboration. So no free-living cell can be half-prokaryote, half-eukaryote. Even if such a cell could have formed it could not have flourished. Apparent exceptions to this rule such as *Giardia* (protists that lack mitochondria) are parasites; they appear to be degenerate eukaryotes.

Sex and death

All single-celled organisms, both eukaryotes and prokaryotes, reproduce by simple division. An individual that replicates by fission cannot be said to "die" if its offspring live on. Individuals *can* die, as a result of a hostile environment or predation, but reproduction by fission affords a kind of "immortality" for the few. When the first multicellular eukaryotes appeared, between 1,500 and 1,000 million years ago, they probably reproduced by shedding parts of themselves. Each portion grew into a new individual genetically identical with the parent. (Spores are a specialised way of doing this.) Again, the genome continued; "immortality" of a sort. This is not true of sexually reproducing organisms. They *all* die.

Why did sex evolve? For a prokaryote with only 1000 or so genes and a very fast replication rate, sex is unnecessary. Even if the mutation rate is high or environmental stresses mount, some progeny will survive. However, a eukaryotic cell with tens of thousands of genes reproduces slowly. In this situation, a high mutation rate or a stressful environment is likely to terminate the lineage. Sex insures against this eventuality in at least two ways.

First, males redistribute genes among otherwise "all-female" lineages, mixing the genomes and generating many variants, giving the species a better chance of surviving environmental change. Males redistribute genes in the gene pool rather as taxation and public spending redistribute money in society. Sex only works *within* species; that is, among organisms with essentially the same genomes. Second, because sexual reproduction fuses the genomes of two individuals, it ensures that every individual has two copies of nearly every gene. As we saw in chapter 11, this insures against potentially damaging mutations.

However, these insurances are long-term. They are advantageous only over many generations. Natural selection operates on *immediate* advantages, not long-term ones. It concerns the current generation, not the future. So how and why did sex evolve? Perhaps it first appeared at a time of high environmental stress, when conditions were changing rapidly. For an organism that contrived to reproduce both asexually and sexually, the sexual offspring – comprising more variants – might have had an advantage over the less varied asexual ones. They might have had a better chance of survival in the changing world.

But this still begs the question of how sex originated. One idea, favoured by Margulis and others, is that it resulted from incomplete cannibalism. A protist devoured another protist of the same species, but the nucleus of the prey survived and fused with that of the predator. Such events can be observed today, so the idea is not implausible. Nevertheless it is a big step from such a coincidental beginning to the formation of specialised reproductive cells, which are found in all sexually reproducing multicellular organisms today. These reproductive cells continue the lineage; the individual that houses them ages and dies.

Sex in humans and some other primates is chromosomal. It depends on whether the individual carries two X chromosomes (female) or one X and one Y (male). The X chromosome has several thousand genes, the Y chromosome only a few dozen. These few dozen include specifically male genes, such as those needed for manufacturing sperm, but the Y chromosome also has 19 genes in common with X. These 19 are found in four groups on the X chromosome. Comparative genetic mappings of several species has shown that chromosomal sex evolved in four stages, the first occurring around 300 million years ago and the most recent around 40 million years ago, when the ape-monkey line parted evolutionary company with the ancestors of the lemurs.

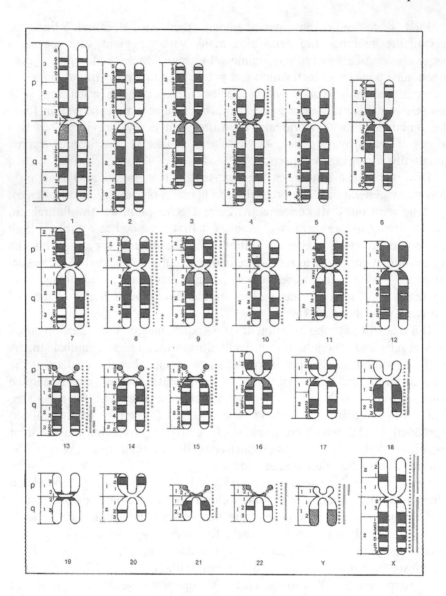

Fig. 13-5: drawing of chromosomes in a dividing human cell, showing the sex chromosomes (X and Y).

In most animals and plants, sex is not fixed by the presence or absence of a Y chromosome. In many species of fish, for instance, one individual in a school becomes male and the rest remain female; if the male dies, another individual becomes male instead. Change the temperature just a few degrees and young salmon develop as females rather than males.

Why did sex originate? Why, in species such as ours, did it become chromosomal? We can only speculate, but attempts to answer these questions have led to new insights. In particular, the discovery that evolution can occur in jumps, as in the evolution of the Y chromosome, shows that it is not always the steady gradual process that Darwin originally envisaged.

One of the most important points about sex is that it allows more complex organisms to flourish: it increases the redundancy of the genome and thereby stabilises it. The more complex the organism, the bigger the genome. The bigger the genome, the higher the risk of fatal damage to the DNA between one generation and the next. The faster the accumulation of DNA damage, the faster the extinction of the species. It is because sex insures against mutational damage that it allows complex organisms to survive. There is a theoretical limit, beyond which any further increase of complexity would over-stretch the DNA repair machinery, but perhaps only organisms as complex as ourselves approach that limit.

Extinctions

The origins of photosynthesis, eukaryotes, sex and multicellularity are probably the most important events in evolutionary history since the origin of life itself. Some people would add the origins of animals, around 600 million years ago, because the familiar structure of food chains, predator and prey, hunter and hunted, dates from this time. The antecedents of animals might have been the curious multicellular organisms of the late Precambrian era, the so-called "Ediacara fauna". Others might add the origins of plants, apparently as a result of symbiosis between algae and fungi around 450 million years ago, because this presaged the spread of life from sea to land. These events have received more coverage in popular books than the topics we have examined in this chapter, because fossil evidence tells us far more about them. But deciding which events are "important" is a matter of viewpoint. From the point of view of the archaea, hardly any of the events listed in this chapter have mattered at all; only photosynthesis was significant for the archaea – and that for largely negative reasons.

From an anthropocentric rather than an archaeacentric stance, the periods of mass extinction were also salient events in life's history. The fossil evidence tells us that the five most recent extinction episodes occurred at intervals of roughly a hundred million years, but it says nothing unequivocal about the causes. The best-known hypothesis blames large asteroid impacts on the Earth. Astronomical data support the view that large asteroid impacts afflict the Earth at roughly hundred-million year intervals. An asteroid of about 10 km diameter almost certainly struck the Earth some 65 million years ago, coinciding with the most recent mass extinction, which ended the age of the dinosaurs: this marked the Cretaceous-Tertiary (CT) boundary. This impact must have raised a massive dust-cloud or a vast amount of water vapour, depending whether the asteroid struck land or sea, so following the immediate devastation there would have been severe global climatic changes that lasted for many years.

There are two difficulties with this hypothesis. First, there is no compelling evidence that asteroid impacts coincided exactly with any mass extinction other than the CT boundary. Second, many of the species that went extinct at the CT boundary disappeared from the fossil record a million or two years *before* the asteroid struck.

An alternative hypothesis hinges on the melting of mantle rocks caused by the collision or fragmentation of tectonic plates. When this happens, the affected parts of the Earth's surface are flooded with molten rock, forming "flood basalts" such as the Deccan Traps of northern India and the Siberian plain. Flood basalt eruptions are not single events but continue sporadically over hundreds of thousands of years of tectonic trauma. They cause gross local destruction of life and release millions of tons of carbon dioxide and sulphur dioxide into the atmosphere. The former gas heats the Earth (the greenhouse effect) and the latter cools it; both lower the atmosphere's oxygen content; and both cause acid rain that poisons land and sea. Flood basalt eruptions coincide with three of the five known major extinctions. The Deccan Traps were formed in the late Cretaceous. The Siberian plain was formed at the end of the Permian, the third and biggest of the known extinctions. The Permian extinction occurred when the ancient super-continent, Pangea, began to break up. The fossil and geological record shows that the sea was acidified, the atmosphere was depleted of oxygen and there was widespread glaciation. All these are plausible consequences of flood basalt eruptions caused by the fracturing of the giant land-mass; collectively, they might account for the loss of 19 out of every 20 extant species that the fossil record indicates. The third known correlation connects a flood basalt plain on either side of what is now the Atlantic with the late Triassic extinction, which occurred about 230 million years ago.

Perhaps no single cause explains every mass extinction. As we suggested earlier in the chapter, oxygen pollution might have caused a mass extinction when photosynthesis first evolved. Man-made pollution is now contributing to another. The rate of species extinction during the past few decades has probably been greater than at any time in the 3,800 million year history of life. The cause this time lies in the poor control of the activities of the most intelligent species of all.

Final extinction

Statistically, it is highly unlikely that the Earth will ever suffer a cometary or asteroid impact sufficiently large to exterminate *all* life, though future serious collisions are to be expected at roughly hundred million year intervals. It is also unlikely that intense radiation from a supernova will ever sterilise the planet, though this is possible in principle. No star close enough to cause total extinction by these means (within 50 light years) is likely to become a supernova during the expected lifetime of the sun.

However, life on Earth will end. As the sun continues to heat up, the buffering effect of the carbon dioxide cycle will ultimately be swamped. The present global warming effect of carbon dioxide production by human activity will have no significance in the long term. Increased temperature will fix more and more carbon dioxide in the form of weathered rock, and although the Gaia effect will mitigate this to some extent - granted the adaptation of future species to changing conditions - the average surface temperature will reach 50°C in some 1,500 million years' time. Only some protists and prokaryotes will survive this temperature. Increased evaporation of water will enhance the greenhouse effect until, in 2,000 million years or so from now, the oceans will boil dry. There is no life as we know it without liquid water.

The last survivors of life on Earth will probably be archaea, organisms not dissimilar to the earliest pioneers. We already know about the archaea that live in deep ocean trenches and require high temperatures to survive. No other known organism will survive when the end of the oceans approaches, and even the archaea might die out when they cease to receive the products of photosynthesis.

But of course we have no way of knowing what species will evolve in the future or what capabilities they will have. When we consider how life on Earth has transformed itself and the planet during the past thousand million years, how can we hope to predict the next thousand million? One thing we can predict fairly safely, though: our own species will not be around for anything like that long.

Chapter 14

THE ORIGIN OF LIFE
Some major ideas and unanswered questions

Life on Earth seems to have evolved continuously for at least 3,800 million years, punctuated by intervals of mass extinction. The very earliest cells were probably archaea-like, tolerant of hot acidic conditions and an oxygen-free atmosphere. Continual changes in DNA generated novel organisms, which were subject to rigorous selection by the environment. Some survived and became established, altering the chemistry of their surroundings. The environment changed. Photosynthesis opened the door to the evolution of oxygen-using organisms and consumers. Symbiosis fused disparate genomes into new and ever more complex living forms. Single-celled eukaryotes emerged, then multicellular organisms, then animals and plants. Sexual reproduction appeared, food chains were born, the land was colonised. As a result, a huge diversity of species inhabits the world today.

Life on Earth has remained astonishingly tenacious. It continues to evolve. Notwithstanding widespread future extinctions, including that of *Homo sapiens*, it will go on evolving until increasing solar energy output boils the oceans dry. But how did it all begin?

The origin of life remains a fascinating and elusive topic. We have even less evidence about it than we have about the origins of photosynthesis or eukaryotes or sex, yet it is the focus of more speculation and debate. It is not a single problem but a constellation of problems, none of which has been fully solved, and different people think of it in different ways. Some authors have focused on the prebiotic formation of organic molecules, or of proteins, or nucleic acids, or membrane-like structures. Others take the view that to explain the origin of life, it is sufficient to explain the origin of molecular self-replication. We suggest that the phrase "origin of life" should denote the formation of a system fitting the characterisation of "livingness" that we established earlier (see the schematic diagram at the start of chapter 10).

The problem of "spontaneous generation"

Every organism is the offspring of previously existing organisms. The evolution of life is a continuous process. Life cannot spring from non-life, as many proponents of "spontaneous generation" believed before the middle of the nineteenth century. Yet it seems that spontaneous generation must have happened at least once; that is what we mean by the "origin of life". Kelvin and other late 19th century luminaries maintained that the origin of life could not have happened by means accessible to scientific knowledge and reason, precisely because it would have entailed spontaneous generation. Such was the influence of these luminaries that the topic did not receive serious scientific attention until well into the 20th century.

The genetic code is more or less universal: all proteins in all organisms are made from the same amino acids, and all amino acids in proteins have the same "handedness" - as do most other biological molecules. Also, all organisms have certain key metabolic pathways in common. This suggests that every organism extant today, and all organisms that lived in the knowable past, can be traced to a single common ancestor[31]. This ultimate ancestor might not have been the first organism – there might have had predecessors that went extinct – but it is hard to deny that it (and any such predecessors) arose from non-living matter, that it was a product of spontaneous generation. If spontaneous generation happened when life began then obviously it was *possible* at that time. But very shortly afterwards it *ceased* to be possible, otherwise we would not be able to trace all organisms to a single common ancestor.

Several attempts have been made to evade this inference. Some authors have suggested that life began elsewhere in the galaxy (presumably within the solar system) and was transported to earth by meteorite or comet, presumably in the form of spores. However, such spores would have had to survive the conditions of interplanetary space, battered by cosmic radiation,

---

[31] An indefinitely large number of amino acids *could*, chemically speaking, be incorporated into proteins, but only twenty actually *are*. The genetic code, the correlation between each amino acid and the DNA/RNA base triplet that encodes it, is constant over all organisms known today - there are minor exceptions only in mitochondrial DNA - and there is no convincing chemical reason why this should be so. Life that arose independently from non-living matter would almost certainly have chemically different proteins and a different genetic code, assuming that it used proteins and nucleic acids at all. Moreover, amino acids and other biological molecules exist in two or more mirror-image forms (*isomers*). These forms are geometrically different but chemically identical, yet only one form – left-handed amino acids, for instance – is used by organisms. The most likely inference is that the choices of amino acids and their nucleic acid correlations were established (by chance) when life began, and have remained fixed ever since. In other words, all extant life has descended from the same ancestor.

for many millenia; and the body transporting them would have reached a very high (sterilising) temperature when it finally accelerated through the atmosphere to the Earth's surface. So this is an unlikely scenario. Even if it were true, it would merely shift the problem of the origin of life to another world, even less well understood than the prebiotic Earth. Therefore, the extraterrestrial origin hypothesis does not answer the question of how life began; it tries to dodge it, and by a rather implausible argument.

Another attempt to evade the problem assumes the steady-state rather than the "big bang" theory of cosmology. If the cosmos has always existed, i.e. had no beginning, then it is possible to suppose that life too has always existed. Therefore, the question of its origin becomes void. This is the position famously adopted by Fred Hoyle and his colleagues. It has found little support during the past few decades because the steady-state theory of cosmology is now almost universally rejected; too much evidence favours the "big bang" alternative. But if Hoyle were correct, we would have to infer that "life", being in effect as eternal and omnipresent as the cosmos itself, is somehow written into the laws of physics. In what sense could this be so? This question is, in effect, the origin-of-life problem in disguise.

Such speculations have minority followings; most people tacitly accept that the "spontaneous generation" problem is real. To explain how life began on Earth from lifeless matter is a huge challenge. To explain why spontaneous generation subsequently became impossible, and has remained impossible ever since, may be equally difficult.

The likelihood of life
Science is about regularities in nature: patterns that can be described, explained and predicted. It deals less comfortably (if at all) with unique, one-off occurrences. We have concrete evidence for life on only one planet – our own. And for reasons given in the previous section, it is arguable that the evolution of all surviving organisms began from a single, unique origin-of-life process. ("Process" means a specific sequence of physical and chemical events. No biologist could believe that life arose from non-life in a single step.) We have no direct evidence about this process and there is no consensus about the sequence of events involved. Everything we can say about the origin of life is a mixture of inference and guesswork. Even our most basic questions about it invite speculation; we are in no position to argue from incontrovertible fact. The best we can do is to temper our speculations with scientifically informed reasoning.

One of the most fundamental questions is this: was the origin of life likely or unlikely? Both alternatives can be, and have been, supported by reasoned (though circumstantial) argument. Broadly speaking, before the closing years of the 20th century, a consensus of scientists believed that the process was astronomically unlikely. Since then, the consensus has perhaps swung the other way. But in this field, fundamental shifts of opinion are matters of fashion rather than advancing knowledge and insight. Both opinions remain legitimate.

First, there are good arguments that the origin of life on Earth was highly improbable. The order of events involved in the process is controversial, but there is broad agreement about what many of the events were. Organic molecules such as amino acids and nucleotide bases had to be present in the right environment. These had to polymerise so that proteins and nucleic acids formed. A dialogue had to be established so that the nucleic acids directed the synthesis of the proteins, and the proteins catalysed the expression and replication of the nucleic acids. These processes had to be enclosed in a membrane-bound system. Energy-providing and signalling apparatus had to be constructed. Organisation had to be imposed. And so on.

All or most of these events are inherently improbable. Few can be simulated in the laboratory and many can only be described in terms that are difficult to relate to the nuts and bolts of physics and chemistry. It is astronomically unlikely that a long sequence of individually improbable events could happen, by chance, in exactly the right order. Therefore, the origin of life was astronomically unlikely. (The probability that all our proteins could have been formed by chance, with the correct amino acid sequences, has been estimated at $10^{-40,000}$, which is effectively zero.) This conclusion has the advantage of circumventing the "spontaneous generation" problem (see the previous section). Explaining why life *stopped* originating once it was established becomes a non-issue: it was simply too unlikely to have happened twice on the same planet.

There are equally strong arguments that the origin of life was probable. First, life began while the Earth was still very young, almost as soon (it seems) as the surface was able to bear liquid water. That is to say: as soon as life became possible, it happened. Second, the conditions on the early Earth probably resembled those in which the deep ocean vent archaea live today. However, the environment was so violent and catastrophe-prone – continual comet and large meteorite impacts, ultraviolet irradiation, incessant volcanic eruptions and earthquakes – that an individual newly-formed organism could have had little chance of surviving and leaving offspring. So it is highly likely that life was extinguished very soon after it began. Presumably, therefore, life must have begun on Earth many times, and only one of the original organisms managed to survive to become our ultimate common ancestor. If so, the origin of life was a common occurrence on the primitive Earth, so it could not have been unlikely. Third, it is now widely believed that life also began in other parts of the Solar

System – on Mars, for instance, and perhaps the Jovian moon Europa – though it might not have survived for long on these bodies. If life began on several planets or large moons in the same Solar System, it could not have been a particularly improbable event.

At present we have no way of deciding between these opinions. If future space probes reveal evidence of past life on, say, Mars or Europa, the "life is probable" position would be favoured. But the opposite view could still be maintained. Perhaps spores of life were, after all, transmitted from body to body in the solar system by meteorites. Failure to find evidence of past life in other parts of the Solar System would not affect the argument at all: absence of evidence is not evidence of absence. Of course, if it were firmly established that extraterrestrial life once existed and had a different molecular basis from ours (a different genetic code, different amino acids, or even alternatives to proteins and nucleic acids), then the "life is probable" option would become almost certain. But that is science fiction.

It is worth noting here that the sequence of DNA bases in the genome of any organism is "random". It is information-rich and unpredictable by any law – it is, in technical language, "algorithmically incompressible". But it is also highly specific; it carries semantic *meaning*, in the sense that it determines and directs the amino acid sequences of all the organism's proteins. No known law of nature allows, still less specifies, the production of "highly specific randomness". Nothing in science tells us how an object with semantic meaning can arise by physical and chemical processes. Evolution by mutation and natural selection produces highly specific randomness, but only by operating on organisms already in existence. This gives us no clues about how the first-ever organisms came to exist. More than one author has suggested that laws of nature crucial for explaining the origin of life remain unknown to us.

The prebiotic Earth; locating the origin of life
After the Earth had formed and its surface had cooled enough for liquid water to accumulate, it remained a very hostile environment from our perspective. Temperatures near the surface must have been around the boiling point of water, volcanoes and earthquakes were ubiquitous and almost everyday occurrences, the atmosphere consisted largely of nitrogen, carbon dioxide and water vapour (there was no oxygen), massive thunderstorms were almost continuous, drastic fluctuations of atmospheric pressure were frequent, and the weak young sun poured ultraviolet irradiation through any breaks in the storm clouds. Worst of all, comets, cometary debris and meteorites battered the young planet continually. On the face of it, conditions seemed appropriate for mass extinction rather than the beginning of life.

Where in this Dante-esque world could the first organisms have acquired a foothold? The "warm little pond" of Darwin's speculation did not exist. The boiling oceans, popularly considered the cradle of life, are implausible candidates: organic molecules in hot dilute solution are far more likely to break down than to assemble into more complex structures, and the oceans were unlikely to favour the coalescence of anything resembling cells. The arid and unstable land surface was riven by volcanoes and comet collisions. Therefore, it seems implausible that life began anywhere on the Earth's surface. This leaves us with candidate locations above or below the surface: (1) in deep ocean hydrothermal vents, (2) in subterranean rocks or (3) in the atmosphere.

The first candidate - hydrothermal vents - has been the most popular since "dark smoker" ecosystems were discovered.   Brands and his colleagues, Russell and Hall and other authors maintain that the earliest organisms resembled archaea rather than bacteria; archaea contain very slowly evolving genes, use energy-producing chemistry consistent with the hydrothermal vent environment and are intolerant of oxygen. These authors point out that hydrothermal vent archaea are chemoautotrophs, taking energy and material sources from their immediate environment, requiring neither sunlight nor atmosphere.   In the hydrothermal vent environment, the formation of organic compounds is thermodynamically favoured. The small channels have a high total surface area suitable for catalysis. However, there are difficulties with this view.   Modern deep-ocean archaea may have an indirect requirement for photosynthetic products falling from the ocean surface; if so, they do not wholly resemble the earliest organisms. Also, it is hard to see how cellular structure could form from molecular components in such a turbulent environment as a hydrothermal vent.   And it is by no means certain that deep ocean vents on the primitive Earth were immune to sterilisation by cometary impacts.   None of these objections is fatal to the hypothesis, but they would all have to be answered convincingly before the idea could attain consensus.

The second candidate – subterranean rocks – has so far gained only minority support.   Thomas Gold believes that the newly-formed planet contained hydrocarbons, the components of natural oil and gas. According to Gold, the world's oil and gas reserves are made of just this material. Received wisdom tells us that oil and gas are the remains of once-living organisms.   Gold says the opposite: the first organisms were made from the hydrocarbons in oil and gas deposits.   There is some evidence for this.   In Australia, oil has been discovered that is 3,000 million years old; it could hardly have formed from fossil organisms.   In Sweden, oil has been found under nearly seven kilometres of rock, below what is usually considered the biosphere.   Helium, a widespread product of radioactive decomposition in rocks, is always found with oil and gas deposits, never on its own. The deep subterranean environment was well protected from comets, meteorites, ultraviolet irradiation, atmospheric changes and other threats. The raw materials of life were available – hydrogen, iron, manganese, sulphur and other elements, in addition to the organic compounds.

Was life forged deep underground from the Earth's primitive organic constituents, only reaching the surface later? Not many people agree with Gold, but so far as the formation of biological *molecules* is concerned, his argument is hard to fault. What is less clear is whether *cells* could have formed in subterranean rocks. However, substantial and thriving populations of archaea have been discovered in rocks several kilometres underground, and they might live independently of the products of photosynthesis. Are these subterranean organisms the direct progeny of the earliest life on Earth? The possibility is exciting because it implies that life originated in an essentially solid medium (rock) rather than a liquid one (sea water), as has usually been supposed. But current opinion is sceptical.

What of the third candidate – that life began in the atmosphere? A few years ago, Tuck and Murphy made the striking observation that the stratosphere holds droplets of ocean water containing up to 50% organic matter. Some of these droplets are remarkably stable; their stability depends on their size. Big ones soon fall, very small ones fuse together, but droplets a micrometre or two in diameter remain suspended for many days. Their organic contents are concentrated by evaporation. These include greasy molecules that cover the droplet surface. If such a droplet falls through a similar greasy surface layer when it re-enters the ocean, the layers will fuse to produce something very like a cell membrane.

Granted that greasy molecules were present in the turbulent oceans of the primitive Earth, this droplet process would certainly have been commonplace. Could the molecules necessary for forming the first organisms have been trapped in high-altitude droplets? Exposure to intense ultraviolet radiation in the upper atmosphere would promote some chemical reactions, though nucleic acids would probably have been damaged. Coincidentally, the most stable suspended atmospheric droplets are almost exactly the size of prokaryotes[32], and with a greasy membrane around them they bear a closer structural resemblance to cells than anything else so far conjectured about the origin of life. This possibility has not been widely considered, but it has no less intrinsic merit than the other alternatives.

---

[32] The stable size depends on gravity and atmospheric pressure. If, say, the same process had taken place on the newly-formed Mars, the stable droplet size would have been considerably smaller. It would have conformed to the size of the deposits found a year or two ago on a certain Mars-derived meteorite recovered from Antarctica, amid much public excitement. Coincidences do happen.

The source of organic molecules

Almost everything about the origin of life is mysterious, but when we turn to the sources of organic chemicals such as amino acids and sugars we have almost an embarrassment of riches. There are broadly three candidates. (1) Organic molecules were made from inorganic ones on the Earth's surface. (2) They were present in the protoplanetary disc from which the Earth formed so they were (and, according to Gold, still are) trapped in the planet's fabric. (3) They were imported by way of meteorite impacts or cometary fragments.

(1) The first suggestion, manufacture from simple inorganic components of the primitive atmosphere, is historically important. The concept of a "primordial soup" in which organic compounds were formed and life originated was first proposed by Haldane in the 1920s, but it awakened scientific interest only in the 1950s, when an attempt was made to simulate prebiotic conditions in the laboratory. In 1952-3, Miller and Urey showed that if electric sparks (simulating lightning) were fired for several days through a gas mixture containing ammonia, hydrogen, carbon dioxide and water (allegedly representing the primitive Earth atmosphere), a tarry mixture formed that contained simple organic compounds such as amino acids and sugars. Miller and Urey were almost certainly wrong about the composition of the atmosphere; in reality it probably contained little or no hydrogen or ammonia, without which no amino acids would have formed in the experiment. As it was, Miller and Urey obtained only a few of the necessary amino acids. Moreover, a dilute solution of organic compounds in the prebiotic sea would hardly give rise to a rich "primaeval soup" from which life could have arisen, as they suggested. The importance of the Miller-Urey experiment is not that it elucidated the origin of life, but that it made it a subject of reputable scientific inquiry. It has retained this status ever since.

The argument behind the experiment was flawed but the conclusion that Miller and Urey drew might be valid. Volcanic vents rich in iron and nickel sulphides could have acted as primitive hydrogen sources, reducing nitrogen to ammonia, so there could have been enough ammonia *locally* in these environments. Amino acids and other organic compounds could therefore have been manufactured from inorganic materials in volcanic vents, particularly hydrothermal vents.

One variant of this idea holds that the organic products of vent reactions became trapped in iron sulphide bubbles, the precursors of cell membranes, the surfaces of which catalysed the formation of protein-like polymers from amino acids. Such bubbles could have formed at the interface between hot alkaline water from the vent and cold acidic sea water. The electrical potential across the iron sulphide membranes could have served as an energy source. This is an attractive possibility because amino acids do not polymerise efficiently if they are simply dissolved in water (though prolonged heating might help them to do so). Efficient polymerisation usually needs a solid surface. On the other hand, the hypothesis implies that proteins appeared before and independently of nucleic acids, which is not a currently popular view (see below).

(2) The second suggestion, that organic molecules were present when the Earth was formed, is supported by astronomical data and is the basis of Gold's idea about where life began (previous section). Infrared telescopy shows that simple organic molecules are widely distributed around the galaxy in interstellar dust: carbon monoxide, formaldehyde, methanol, polyaromatic hydrocarbons, and some amino acids. Some meteorites - a type known as carbonaceous chondrites - contain organic compounds including amino acids. Since these meteorites are believed to be remnants of the protoplanetary disc of the Solar System, the primitive Earth probably contained the same compounds.

(3) The third suggestion, that organic matter reached the primitive Earth surface via meteorite and comet impacts, is almost certainly true. It is supported by the same astronomical data as (2). Many comets are rich in simple organic compounds. The comet storms that scarred the planet in its youth were probably the source of most of the Earth's water (water is the main ingredient of most comets) so they could have been a major source of organic compounds as well. Even today, when impacts are very much rarer, some 50,000 tons of meteorite dust fall on the Earth every year, and this too contains traces of organic constituents[33].

Whatever their source, it seems clear that simple organic molecules, the raw ingredients of life, were abundant on the prebiotic Earth. Most of them arrived ready formed, either native to the planet or delivered by meteorites. Some might have been made by Miller-Urey processes in such environments as volcanic vents. The provision of simple organic constituents is one facet of the origin of life that no longer seems problematic.

The molecular chicken and egg problem

The early Earth was well provided with simple organic molecules, but how were these turned into proteins and nucleic acids? How they were *polymerised*? This remains unanswered. There is an even more contentious question: which came first, nucleic acids or proteins – or, perhaps, cell

---

[33] This sounds a lot, but the Earth's surface is big. Fifty thousand tons a year works out to about three micrograms (three millionths of a gram) per square metre every day.

membranes? Each of these possibilities has been championed. Each entails considerable difficulties.

During the period between 1952 and the early 1980s, debates about the origin of life often took the following form. If the first *big* organic molecules made on Earth were proteins, how did they replicate? Proteins are not normally self-replicating (there are exceptions, but they are very special cases). Could they have given rise to nucleic acids that encoded them? If so, how? How did polymers of exclusively "left-handed" amino acids form, and how (since random choice would have been many orders of magnitude too inefficient) did *meaningful* sequences of amino acids arise, producing functional proteins? And how could proteins and nucleic acids have been held together in the same confined space, so that replication and translation became coherent? On the other hand, if nucleic acids came first, how were they replicated with no protein to act as a replicating enzyme? How were the bases aligned to form *meaningful* sequences? And once again, how - when the replicating enzyme finally appeared – did they become confined in the same small space? Finally: if the membrane (the confiner of the space) came first, what was it made of? How did it replicate itself? How did it acquire proteins and nucleic acids to replicate inside it?

Turning amino acids into proteins requires a good deal of heat and (normally) a solid surface to act as catalyst. The iron sulphide deposits of hydrothermal vents mentioned in the previous section might have sufficed. But only left-handed amino acids are found in proteins. Therefore, the prebiotic proteins that contributed to the origin of life must presumably have been "left-handed". Would iron sulphide catalysis have been so selective as to polymerise only left-handed amino acids, ignoring the right-handed ones? There is no evidence that it would. What alternative is there? *Could* proteins have been made without nucleic acids?

At least some nucleotide bases can be made under Miller-Urey conditions, and there was plenty of phosphate on the primitive Earth, so given a prebiotic source of the appropriate sugar (ribose), most ingredients of nucleic acids were available. Quite how base-sugar-phosphate units were assembled and then polymerised is a matter of conjecture, but it happened somehow. It is generally agreed that the first nucleic acids were RNA-like not DNA-like. DNA is chemically more exotic and it probably entered the scene later. But how was it possible to assemble RNAs of reasonable size under prebiotic conditions? Were solid-state catalysts again involved? How was the correct "handedness" imposed? Could RNA have been made without proteins?

At about the time of Haldane's "primaeval soup" conjecture (the 1920s), Aleksander Oparin proposed that the earliest proto-organisms were membrane-bound globules that accumulated ingredients from the environment and "replicated" by random fission. Oparin found that when glucose, a starch-making enzyme, gum arabic and histones were mixed together in solution, self-replicating globules formed. These "coacervates" suggested that membranes might have formed spontaneously and become self-replicating under the right circumstances. The suggestion (see above) that life originated from membrane-bound droplets in the atmosphere is a modern version of Oparin's conjecture and it has circumstantial support. But the questions remain: how and where did such droplets become filled with proteins and nucleic acids, and how and where were those polymers produced?

The "RNA world"
This chicken-and-egg debate was transformed during the 1980s by a novel discovery: Cech and Altman found that RNA can function as an enzyme - no proteins were necessary. Some RNAs catalyse their own cleavage and their own polymerisation. Orgel had demonstrated some years previously that polynucleotides, particularly RNAs, can catalyse the formation of copies of themselves. These strands of evidence suggested to some scientists that the question "proteins first or nucleic acids first?" was answered: the nucleic acid, specifically RNA, came first. Gilbert and others proposed that for a period on the primitive Earth, RNA molecules manipulated themselves and each other, replicating autonomously. They called this period the "RNA world".

The RNA world hypothesis is now textbook material. There is a consensus that the "RNA world" gave way to "true" life when the RNA started to translate itself into proteins and DNA succeeded it as the repository of genetic information[34]. But the hypothesis is inadequate. First, it is hard to make key RNA reactions go without external catalysts, and bringing the four bases together for initial synthesis would have been problematic. Second, RNA molecules are fragile and tend to break up unless

---

[34] DNA is chemically much more stable than RNA and is far less inclined to catalyse reactions that will alter it. It is more suitable for making very big polymers that are more or less guaranteed to last. Natural selection would certainly have favoured the replacement of RNA by DNA as genetic material, but it is still far from clear how DNA ever got into the act in the first place.

carefully cosseted. The longer the RNA, the more fragile it is; but a short RNA is relatively useless both as a repository of information and as an enzyme. Perhaps several short RNA molecules working together could have formed a replicating system, but assembly of a number of short RNAs in the same confined space would have been improbable. Third, RNA seems unable to catalyse many of the reactions crucial for energy metabolism. Fourth, as in the case of amino acid selection for protein manufacture, it is not clear how nucleotides with the correct "handedness" were selected for polymerisation. Fifth, there are major differences in the RNA replication mechanisms of archaea, bacteria and eukaryotes, suggesting that these mechanisms had no common ancestry. This throws doubt on the idea that all major branches of life arose from an "RNA world". Finally, it is not clear how the proto-organisms of the "RNA world" were supposed to manage without membranes. There is no indication of internal state, or responsiveness to environment: autonomously replicating RNA, if it ever existed outside a modern laboratory, was not an organism.

Of course, prebiotic "RNA" might have contained ingredients that modern RNA lacks. Some chemical modifications confer remarkable chemical properties on the molecule (as they do on DNA). When bases are modified or novel ones are inserted, or a chain of amino acids is attached to the end of the polymer, RNA can perform or catalyse all kinds of reactions, perhaps including some that simulate energy metabolism. An RNA-protein hybrid might have forged a link between the replication and translation processes we know today. Present-day living cells cannot be persuaded to accept these modified nucleic acids, but that does not mean they played no part in the *origin* of life. On the other hand, there is no evidence that they did. Modified nucleic acids are interesting to chemists and might be valuable commercially, for instance in manufacturing certain drugs, but whether they make the "RNA world" more plausible is dubious.

Other self-replicating chemical systems have been studied. One of the most interesting discoveries in this field is due to Julius Rebek. He found that when a self-replicating polymer is mixed with inefficiently replicating polymers, together with their building blocks, the best replicators quickly predominate at the expense of their competitors. This result seems obvious with hindsight; a kind of chemical Darwinism. But it might help us to understand how self-replicating chemical systems came to be "selected for" on the primitive Earth. Could an RNA-protein hybrid have developed by such a mechanism? This is speculative, but it is not impossible.

Inorganic proto-life?

Which came first, the living state or the molecules (proteins, nucleic acids and membrane components) on which it depends? Nearly everyone would say that the molecules came first and the living state somehow arose from them. But alternative is logically possible. The first living things might have been made of entirely different materials, which were subsequently replaced by nucleic acids and proteins.

Cairns-Smith drew attention more than 30 years ago to the "replicating" properties of certain kinds of clay. Kaolinite, for example, forms large flattish crystals that stack like playing cards. Defects formed in a kaolinite crystal when one atom is replaced by another can be replicated: a new crystal formed adjacent to it repeats the defect, maintaining the same overall shape. Thus, kaolinite behaves superficially like DNA. It replicates itself, so long as it is supplied with the right ingredients; and it passes on its acquired defects to subsequent "generations". However, although kaolinite's crystal structure is quite complicated by mineral standards, it is incomparably simpler than DNA and therefore far more likely to form by ordinary physical and chemical processes. It was probably common on the prebiotic Earth.

Cairns-Smith suggests that life began as a replicating clay mineral system. Many clay minerals bind organic molecules such as nucleic acid bases and amino acids, which profoundly alter their properties. Montmorillonite, for example, becomes soft and pliant in the presence of some organic compounds but hard and brittle in the presence of others. At the same time, the clays catalyse reactions among the bound organic molecules. Under some conditions, they might catalyse polymerisation. Over millions of years, suggests Cairns-Smith, organic components attached to the replicating mineral system became steadily more complex and played an increasingly important part in the replication process, until they replaced the original clay mineral altogether. He calls this hypothesis "genetic takeover".

The idea is attractive for a number of reasons. It deals directly with the most profound question of all: how did undirected physico-chemical processes generate the inevitably high complexity and "semantic content" of the first living organism? It places the origin of life in a solid rather than a liquid environment, which is consistent with the production of order – and cells are certainly ordered. The minerals required were probably abundant on the primitive Earth. The hypothesis is chemically plausible. Some commentators have observed that Martian dust clouds are rich in Montmorillonite. Most of all, the idea is attractive because it encourages us to separate the question of life's origins from the chemistry of life as we know it today. We are not obliged to accept the intuitive "molecules first, cells after" approach.

However, nothing in Cairns-Smith's work suggests that his clay systems could have come close to a living state as we characterised it in chapter 10. Could a structure of kaolinite or Montmorillonite (with or without organic additives) have had reciprocally dependent internal structure, metabolism, and internal transport processes - an internal state - however rudimentary? Could it have responded in organised ways to external stimuli or exhibited anything analogous to control of gene expression? Even if the answers are "No" (which they probably are), the Cairns-Smith model might still help to explain how replicating nucleic acids came into being; but unless the answers are "yes", we cannot regard the proposed clay mineral structures as "living".

What was the origin of the "internal state"?

Among the authors who have directed attention to the origins of metabolism as well as genes, proteins and self-replication, Freeman Dyson is perhaps the best known. He suggested that membrane-bound structures arising from something akin to the Oparin mechanism might have been of different kinds. Some might have contained replicating equipment, others metabolic equipment. Symbiosis between the two kinds could have produced a primitive cell. But the origin of the "metabolisers" is not clear.

Prigorgine popularised the phrase "self-organising complexity" and noted that autonomous, self-maintaining complex systems adopt stable states that are far from thermodynamic equilibrium. A living cell is just such a system. Kauffman showed that a chemical mixture in which a few components catalyse reactions among the others becomes self-stabilising, self-organising and in the mathematical sense complex. A self-stabilising, autocatalytic chemical system contains the rudiments of metabolism, internal structure and homeostasis – though without genes or replication. This could in principle explain the origin of Freeman Dyson's "metabolisers". The difficulties with this abstract approach are (a) that it lacks experimental support (apart from computer simulation) and (b) that metabolism is not *self-*organised; its organisation is directed by gene expression and signals from the environment.

It is difficult to see how Kauffman's scheme could be related to the RNA world hypothesis or most other conjectures about the origin of life. However, could a Cairns-Smith "clay matrix" have acquired a suitably large array of organic additives for an autocatalytic system to develop? If so, then affirmative answers might be given to the previous questions, and the Cairns-Smith model might suggest a practical way of realising Kauffman's mathematical scheme for the origin of life.

Why did life on Earth stop originating?

If conditions were initially favourable for the origin of life, why did they cease to be so and why did only one lineage ultimately survive? First, conditions must have changed fairly rapidly as the planet and the rest of the solar system settled down. Comet and meteorite impacts would have become less frequent; the "late heavy bombardment" has been dated to 4,000-3,800 million years ago. Second, the Gaia principle tells us that when early life became established it altered the environment. Perhaps, therefore, life itself made the planet unsuitable for the origin of life. Photosynthesis might have been a key factor; even a trace of oxygen in the atmosphere could have permanently sterilised the inanimate world. Alternatively, perhaps one type of cell - our ultimate ancestor - ate the others.

It is easy to dismiss this aspect of the origin-of-life problem as relatively trivial. But as we argued earlier in the chapter, it is trivial only if the origin of life was a highly improbable process. If the origin of life was likely, in other words recurrent, then the question of why spontaneous generation ceased demands a convincing answer.

Final word

The origin of life is a topic that strains the boundaries of science, for reasons we outlined earlier in the chapter. There is far more speculation and argument than evidence. Some readers might therefore consider the subject unworthy of serious consideration.

However, a comment by Morowitz is worth quoting: all today's cells in all today's organisms are genetic and metabolic fossils of the earliest life. The way we are made carries the stamp of our origin. Life on Earth has memory and it remembers its birth. If we are to understand life completely, we need to understand its origin.

Moreover, investigating the origin of life has produced interesting and provocative ideas. No matter how sceptical we might be about (for example) the Miller-Urey experiment and the RNA world hypothesis, these and other contributions to the field have stimulated sound scientific work that has yielded useful knowledge in chemistry and other fields. Many debates about the origin of life have brought together information from astronomy, geology, biology and chemistry in novel and informative ways. This synthesis would not have happened otherwise. And as we said in the first chapter of this book, ideas are enjoyable in themselves if we can hone them by rational debate; particularly when the ideas concern such an intrinsically fascinating topic as the origin of life.

But the ideas that have blossomed indicate that most of our thinking on the subject has progressed significantly only at the *molecular* level. We seem to be as far as we ever were from understanding how the collective structure of the cell came into being. In this respect, Prigorgine's and Kauffman's ideas hold a special place in the literature. If they can be combined productively with (say) the Cairns-Smith model, as we speculated, then there might be truly radical progress in our understanding of the origin of life.

Chapter 15

OTHER WORLDS
The possibility of extraterrestrial life

There are many debates about the origin of life on Earth, but whether there is life elsewhere in the universe is a still more contentious topic, a minefield of unanswered questions. We considered one of these in chapter 14: how likely or unlikely was the "origin of life"? Another basic question is whether all life in the universe is based on nucleic acids and proteins, like ours, or whether it can have different molecular or physical hardware. If so, what are these alternatives? Even if we could answer these questions, we would then have to ask how many planets in the universe could support life, and estimate how many actually do so. Has all extraterrestrial life (if it exists) evolved as ours has? Perhaps most of all, people want to know whether there are intelligent species on other worlds, and if so whether we could communicate with them.

There is scant evidence to help answer these questions, but people nevertheless speculate and even hold firm opinions about them. This has generated an entire discipline, "exobiology" or "bioastronomy", complete with dedicated international conferences and publications. Exobiology is probably the only "scientific" subject ever to have thrived in the virtual absence of data. There is no denying its fascination.

The Drake equation
What do we need to know in order to decide whether life exists in other parts of the universe? This is a reasonable question. How might we locate another technologically advanced civilisation? This might not be nearly so reasonable. To assume that "life" implies "technologically advanced civilisation" is to beg a lot of questions. Nevertheless, considerable effort goes into scanning the heavens for radio emissions that might betoken the existence of one or more technologically advanced civilisations in other parts of the galaxy.

In 1961, Frank Drake, a radio astronomer who later became chairman of the SETI Institute, tried to specify the factors involved in the development of a technologically advanced alien civilisation. He encapsulated his ideas in an equation that has guided much subsequent discussion of the subject. The equation can be written:-

$$N = R^* \times f_p \times n_e \times f_l \times f_i \times f_c \times L$$

where N = the number of civilisations in the galaxy from which electromagnetic emissions are detectable; R^* = the rate of formation of stars that are compatible with the development of life; f_p = the fraction of those stars that have planetary systems; n_e = the number of planets with life-supporting environments orbiting each of these stars; f_l = the fraction of these planets on which life appears; f_i = the fraction of life-bearing planets on which intelligent life evolves; f_c = the fraction of civilisations that develop an advanced technology, emitting detectable radio signals; and L = the length of time during which these civilisations release detectable signals.

The National Research Council of the USA is just one official body that assumes the Drake equation to be a valid guide to research. The equation has the merit of being simple and dimensionally correct, and it focuses our curiosity and interest in a fascinating question. However, it presupposes that intelligent life and the development of advanced technology are natural, inevitable outcomes of some cosmic process, and as we shall see, that is a very dubious supposition. Also, it overlooks a number of important factors.

We shall return to the Drake equation and its connotations at the end of the present chapter. In the interim we shall try to assess the likelihood of life on other planets, using arguments based on earlier parts of this book. This will enable us to identify what is missing from the Drake equation, or questionably assumed in it.

Signs of life

Suppose we travel to a planet in a distant part of the universe. There we encounter an object that might be deemed "living". Let us call it Z. How can we decide whether Z is living? What might have suggested to us that it is an organism?

First, how big or small might Z be? No terrestrial organism has linear dimensions less than about one micrometre. This lower limit is set by the minimal equipment needed to maintain a living state (chapter 2). Organisms with the same chemistry as ours presumably cannot be much smaller, no matter where in the universe they live. Setting an upper size limit is more difficult. Mechanical restrictions give us some guidance - for example, animals with chitinous exoskeletons, such as insects, cannot grow beyond a certain size – but this is not sufficient. Apart from dramatic organisms such as *Wellingtonia* or *Gigantosaurus*, the Earth boasts colossal fungi, with hyphae spreading over almost a square kilometre of soil. The maximum possible sizes of organisms on other planets are beyond conjecture. However, alien organisms such as Z would presumably have cellular structures. We discussed earlier why the range of terrestrial cell sizes is so narrow (chapter 3 and following); we can extrapolate this argument, at least provisionally, to other worlds. No matter how big Z might be, we could (in principle) investigate whether it comprised cells with complex internal organisation.   In any case, the simplest organisms (corresponding to prokaryotes on Earth) are highly likely to predominate on any life-bearing planet, so Z is statistically likely to be microscopic and to comprise a single cell.

This reasoning is tenuous and might not apply to life with a different chemistry. For instance, what about the possibility of organisms based on silicon or perhaps phosphorus instead of carbon? Compounds of these elements produce interesting polymers at high temperatures and pressures. Such possibilities allow no way of predicting minimum organism size. We cannot even be sure whether such hypothetical organisms would have cellular structures. Nevertheless, Gold has suggested that there might be silicon-based life in subterranean parts of the Earth, the "deep hot biosphere". If he were right, could we recognise them as living? For example, life as we know it cannot exist without liquid water, but would water be a prerequisite for silicon-based life?

Can we apply our criteria of "livingness" to objects such as Z on the hypothetical alien world? In principle we can, but there might be practical difficulties. For instance, organised structural complexity is an inevitable feature of life. However, in order to identify "structural complexity", we have to know what sort of structure and what sort of complexity to look for in Z. This entails judgement.   How complex is complex? What objective criteria of "organisation" can we apply? Another aspect of "livingness" is metabolism - the exchange of energy and materials between organism and environment, on which the integrity of structure depends. If Z exchanges energy and material with its surroundings we might be expected to recognise the fact; nevertheless, the inputs and outputs might happen too quickly or too slowly for us to interpret them correctly. Such exchanges might be mistaken for inanimate processes. Would a mineral that (say) absorbed ultraviolet radiation and emitted heat, maintaining a highly elaborate crystal structure in the process, be regarded as an organism that ate ultraviolet and excreted heat, or as a lump of rock with a complicated chemistry? The other hallmark of internal state - internal transport - would never be observed in practice unless we had already decided that Z was an organism. Control of gene expression would be definable only if we knew what constituted "genes" in Z and what we meant by "expression". Responses to stimuli might be noticeable - but once again, only if the time-scale was neither too slow nor too fast. And what if the stimulus was elusive, such as a narrow wave-band in the far infrared - would we ever think of looking there?

Perhaps this is too pessimistic a view. If there were any chance that Z was alive, then surely we would examine it exhaustively against each of our criteria, making proper allowance for all the aforementioned difficulties. Irrespective of complications, a proper examination should suffice to reveal any reciprocal dependences among the aspects of internal state (structure, metabolism and transport) and among internal state, gene expression and responses to stimuli. According to our characterisation, the essence of "livingness" lies in these reciprocal dependences.

Finally, let us remember that no organism is an island; that all organisms belong to ecosystems. Z would not exist in isolation if it were alive. It would be one of many objects on the planet that might be deemed "alive", some similar Z itself, others perhaps different in outward appearance. All these objects would, if we studied them for sufficiently long and in the right ways, show a measure of mutual dependence. Overall, it seems that we could rely on our criteria of "livingness" to make a reasonable decision about the status of Z.

Extraterrestrial life in the solar system
The possibility that life began on Mars as well as Earth has long been entertained. Evidence that there was once liquid water on Mars, that the atmosphere was once denser and the climate warmer, has made many scientists take the idea seriously. Mars is almost certainly dead now - its tenuous atmosphere shows chemical equilibrium, no Gaia effect - but there might have been life there long ago.

In 1996, a meteorite found in Antarctica was heralded as evidence that life had existed on Mars at around the time it began on Earth. The meteorite, weighing less than two kilograms, had cracks containing carbonate deposits, within which were embedded tiny hair-like structures rich in complicated hydrocarbons. These hydrocarbon-rich structures had apparently formed before the rock left Mars. Were they bacteria?

The meteorite bore the name ALH84001, indicating that it was the first such meteorite (001) to have been found in the Allan Hills area of Antarctica (ALH) in 1984. It was known to have originated on a planet because it was mainly igneous; planets have volcanoes but asteroids and comets do not. There was compelling evidence that it came from Mars rather than any other planet. For example, the nitrogen, argon and carbon dioxide contents of gas bubbles trapped in the rock closely resembled those in the Martian atmosphere, and the rock itself contained iron disulphide, quite a hallmark of

Martian origin.   However, some media reports in 1996 gave confused accounts of dating, obscuring the arguments about the meteorite.

Three dates are relevant: the formation of the rock; the formation of the carbonate deposits that caused all the excitement; and the ejection of the rock from Mars in meteorite form.   Rubidium isotope decay evidence showed that the rock itself dated from some 4.5 thousand million years ago, when the Solar System planets were forming.   So ALH84001 was a piece of the original stuff of Mars.   Potassium-argon dating showed that the cracks in the meteorite, possibly consequences of a comet impact on the young Mars, were around four thousand million years old.   The carbonate deposits in these cracks were formed considerably later, no more than 3.5 thousand million years ago.   (This figure was imprecise; the carbonate might be only half or even a third of that age.)   To determine the ejection date, investigators took advantage of the fact that cosmic radiation would have produced carbon-14 while the rock was in interplanetary space.   When the rock reached the Earth this production would have stopped and the carbon isotope would have begun to decay.   Carbon dating indicated that ALH84001 left Mars only about 13-14,000 years ago.

The hair-like formations in the carbonate deposits appeared in the electron microscope as strings of tiny beads about 25 nanometres across. They were surrounded by iron sulphide and magnetite deposits, which are characteristic of the activities of some prokaryotes on Earth.   They contained organic material, mainly polyaromatic hydrocarbons.   These components were not the results of terrestrial contamination: their concentrations were greater in the middle of the meteorite than at the periphery.   (If they had been contaminants, the higher concentrations would have been on the outside not the inside.)   The suggestion that the 25-nanometre particles might be the remains of tiny bacteria, evidence of ancient life on Mars, made media headlines around the world.

They were not fossil bacteria, because they were well below the minimum size of a living cell - around 100 times smaller than a prokaryote (unless, of course, life on Mars was organised quite differently from ours). Moreover, the carbonate deposition seems to have taken place at a temperature around $650^{\circ}C$, incompatible with the survival of even the toughest archaea. This might seem disappointing, but the organic deposits are still interesting. The presence of complicated hydrocarbons on primitive Mars is exciting in itself, particularly when we recall Gold's idea about the origins of oil and gas on Earth (see chapter 14).   But the rumours were premature. There might once have been life on Mars, but ALH84001 was not evidence of it.

The search for signs of ancient life on Mars is still worth pursuing. It might throw light on the origins of the ALH84001 hydrocarbons, and that would be informative. There are also plans to send probes to Europa, Jupiter's largest moon, which might contain liquid water - the surface is covered with thick sheets of water ice rich in organic materials, and the lower depths might be heated and melted by volcanoes. The presence of water and organic material suggests that life with much the same chemistry as ours might have originated on Europa.

If either the Mars or the Europa quest is successful then some of our judgements about the origin of life on Earth might need to be revised. For example, in chapter 14 we considered the possibility that terrestrial life was imported by meteorite, having begun on another planet. We dismissed this idea is implausible – and as unhelpful, because it fails to address the "origin" problem. However, some authors have argued in favour of the idea, and optimistic interpretations of the ALH84001 evidence were recruited in their support. If clear evidence of life on Mars, Europa or elsewhere is ever found, we might be obliged to reconsider our position on this topic; but at present, we are not[35].

Other solar systems: the supply of planets
Thanks to refined astronomical techniques we now have clear evidence that there are planets elsewhere in our galaxy. Some of these planets are "free" in interstellar space, occurring either singly or in clusters. By 2004, rather more than a hundred had been discovered in orbit around nearby stars. "Free" planets are very unlikely to house life because their surface temperatures can be only a few degrees above absolute zero, far too cold for the complicated physical and chemical processes that must be required for any imaginable form of life. Planets that orbit stars are more plausible candidates.

There is no firm definition of "planet", but although this is a significant issue in astronomy it need not impinge on our discussion here. At the lower end of the size range there is no absolute distinction between planet, planetessimal and asteroid; these are names for different portions of a size continuum. At the upper end, there is no clear way of discriminating between planets and superplanets, or superplanets and brown dwarfs. (A brown dwarf is a very small star that emits no light.) However, asteroids are unlikely to house the components necessary for any kind of life, and the

---

[35] There are stimulating and scholarly books, for instance those by Francis Crick and Paul Davies, that argue powerfully for the position opposed to ours. Many of these works are easily accessible and we urge the interested reader to consult them (see "Further Reading").

same applies to large planets comparable to the gas giants of our Solar System. Gas giants consist very largely of hydrogen, which is no basis for forming complex molecules of any sort. Also, their gravitational fields are huge. So far as we can suppose, therefore, a planet must be substantially bigger than an asteroid and substantially smaller than a gas giant to be potentially life-bearing.

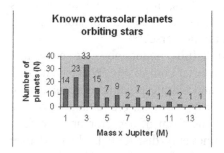 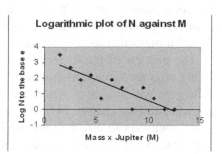

Fig. 15-1: numbers of extrasolar planets known in 2004. In the left hand diagram, the numbers of extrasolar planets orbiting stars that had been discovered before August 2004 are categorised according to their relative masses. Each unit on the horizontal axis corresponds to the mass of Jupiter. Taking account of technical limitations, the two lowest (left-hand) values are probably underestimates. These two values are therefore ignored in the right hand diagram, where the natural logarithm of planet number is plotted against mass. This plot assumes the simplest mathematical model (exponential decay) to fit the data. The fit is imperfect (the correlation coefficient is –0.89), but the data do not justify a more sophisticated model. Given the gradient of the best straight line (-0.272) and taking "Earth-sized planets" to have masses in the range 1/250-1/400 that of Jupiter, the model predicts that the extrasolar planetary systems so far observed contain around 70 "Earth-sized planets". Observation of such planets is beyond the limitations of current astronomical methods.

Extrapolating from limited data is dangerous, but it is interesting to plot the numbers of known extra-solar planets against planetary mass. Because of the limitations of current methods (measurement of dynamic effects, transits, photometric variables, microlensing and so on), relatively low-mass planets have a greater chance of being overlooked than larger ones. Therefore, the numbers at the lower (left-hand) end of the mass scale on Fig. 15-1 are likely to be underestimates. Allowing for this, the data approximately fit an exponential decay curve. If this curve is extrapolated backwards, then for every 56 planets in the range 0.5-2.0 Jupiter masses (the number presently known) we predict 70-80 roughly Earth-sized planets. (Jupiter has about 300 times the mass of the Earth.) This ratio resembles that in our own Solar System, which contains five small planets and four gas giants[36]. If this prediction is valid, then roughly Earth-sized planets are fairly common in those parts of the galaxy that have so far been studied; and by inference, common throughout the galaxy and presumably other galaxies. However, this is a questionable calculation: current techniques are incapable of detecting planets smaller than 6-8 earth masses, and Fig. 15-1 extrapolates from very limited evidence. But suppose the inference were broadly correct. Would it mean that life is widespread in the universe?

The process of planet formation is not fully understood. It seems that young stars are formed from clouds of material comprising 99% gas, mainly hydrogen, with traces of water ice, dust and simple organic molecules such as methanol (see chapter 14). When a star has formed, the remainder of this cloud continues to circulate around it in the form of a protoplanetary disc. The material of this disc is (probably) unevenly distributed. In its denser parts it forms lumps that grow by gravitational accretion. The bigger the lump becomes, the more of the surrounding disc material it draws into itself. The growth is self-limiting because the lump - the nascent planet - gradually clears the neighbouring space of matter until there is nothing left for it to attract[37].

Beyond the "snow line" of the developing solar system, that is, beyond the point at which radiation from the star can keep water above its freezing point, ice in the growing lump might foster the formation of giant planets. This might have been the case in our own Solar System; the four planets nearest the sun are small, but the next four, beyond the "snow line", are gas giants. Organic molecules might help to "glue" the rubble of a growing planet together. But this account is incomplete. Some extra-solar giant planets are very close to their stars, so close that their orbital transits take only a few days, so ice cannot have fostered their formation. Such "hot Jupiters" would swallow a nearby Earth-sized planet as a powerful vacuum cleaner might swallow a baby gerbil.

In our Solar System the planetary orbits are almost circular. In other systems this is not the case. Some large extra-solar planets have orbits that

---

[36] We have included Pluto in our count of small planets, but in fact the Kuiper Belt beyond the orbit of Neptune contains about 100,000 mini-planets (planetessimals), of which Pluto is just one relatively large example.

[37] After planets have formed, the outer parts of the disc might remain. Even beyond the Kuiper Belt on the outside of our own solar system lies the Oort Cloud, the remnant of our protoplanetary disc. It is in the Oort Cloud that comets are formed.

are markedly elliptical and highly eccentric. Smaller planets close to any part of such an orbit would be wiped out. An Earth-sized planet with such an eccentric elliptical orbit would, even if it survived destruction by a gas giant, suffer extreme surface temperature fluctuations during the course of its "year", well beyond the buffering capacity of any Gaia effect. A planet like that is very unlikely to support life, probably regardless of the chemistry involved.

Many stars are large and short-lived. Others are small and prone to storms of high-energy radiation that would eliminate life on any nearby planet: life as we know it, certainly; other forms of life, probably. But there are plenty of stable sun-sized stars in our galaxy and probably elsewhere in the universe. However, even if most of these have orbiting planets, how many of those planets have the right size and chemistry for developing life, and how many of *these* have near-circular orbits, and avoid destruction by gas giants? We have no sound basis for answers, but judging from the limited evidence we have and the foregoing reasoning, potentially life-bearing planets would seem to be quite rare.

There is another difficulty: how important was the moon for the development of life on Earth? Having a single relatively large satellite has helped to stabilise the orientation of the Earth's axis for long periods of time, inhibiting rapid fluctuations of climate. Mars was not so lucky, one likely reason why Mars is now sterile. How many approximately Earth-sized planets describing near-circular orbits around stable sun-sized stars, free of gas giant interference, boast a large stabilising satellite? Given the extraordinary circumstances under which the Earth is believed to have acquired its moon, the answer seems likely to be a low number. Therefore, potentially life-bearing planets are probably rare.

The likelihood of life elsewhere
Granted at least one planet somewhere in the galaxy (apart from our own) that meets the necessary criteria, how likely is it that life *did* begin there - and continued to flourish and evolve? This brings us back to the likelihood of the origin of life, which we discussed in chapter 14. If life on Earth was highly improbable then it is equally improbable everywhere. In other words it has probably not happened, since - as we have just argued - the supply of suitable planets is likely to be limited. Conversely, if there is or has been life elsewhere in the universe, then the origin of life on Earth must have been probable. Therefore, to consider the origin of life on Earth unlikely is, logically, to doubt the probability of life elsewhere in the universe. However, if we accept that extraterrestrial life is likely, then we must also believe that life on Earth was probable. If life on Earth was probable then it

originated several times, only for the process suddenly to become impossible and for all but one lineage of organisms to be wiped out. For this reason, the search for extraterrestrial life, life with an origin distinct from Earth's, is justified: it throws light on one of the major issues that we raised In chapter 14. (Strictly speaking, this argument is valid only for protein and nucleic acid based life similar to ours. We cannot extend the reasoning to life with a completely alien chemistry - life without proteins and nucleic acids, say, or even without carbon. Since we cannot imagine such an alien biochemistry, we cannot conjecture what the conditions for it might be or how likely those conditions are.)

Suppose there is life on planets circulating other stars, and suppose it is protein and nucleic acid based. Will it evolve in a way comparable to ours? Genetic change and natural selection would surely be inevitable, so evolution must occur. But would ever-greater complexity be inevitable, as it seems to be on Earth? Our answer depends partly on whether we side with Kauffman or Gould: does evolution select among a limited though changing choice of patterns (Kauffman) or is it wholly contingent, patternless, a matter of pure luck (Gould)?

A case can be made in favour of progressive increase in complexity: in any developed ecosystem, extreme interdependence is highly likely among at least some species pairs; so symbiosis will probably result, leading to more complex organisms (chapter 13). But this is not a watertight argument. If symbiosis does *not* happen, would more sophisticated organisation still emerge because of the mathematical properties of complex adapative systems, fuelling an evolutionary trend towards increasingly complex organisms? It is not easy to imagine a plausible biological mechanism for this, other than symbiosis. Therefore, the emergence of organisms comparable to eukaryotes, and the subsequent emergence of multicellularity, is a possible but not inevitable (or even very likely) scenario for other worlds. If there is life like ours elsewhere, it has probably remained a prokaryotic enterprise.

Again, we cannot extrapolate this argument to life with an alien chemistry because it might have no analogues of "prokaryote", "eukaryote", "multicellularity", and so forth. The very notion of "evolution" might have to be revised in that situation. The bounds of reasonable speculation are exceeded.

Aside: public attitudes to extraterrestrial life
The story of ALH84001 indicates how we might react to future reports of extraterrestrial life. When the news about the meteorite broke, UFO

enthusiasts responded predictably[38], showing how ready some people are to make the illicit progression: "If there was life, there must have been human intelligence, so there must have been advanced technology". These enthusiasts made the customary references to Martian canals, pyramid-like structures, and the appearance of a giant face on the surface of the Red Planet - all of which are demonstrably the products of imagination. "This discovery is not really a surprise," said the editor of *UFO* magazine, Graham Birdsall. But he voiced a general excitement at the supposed discovery of life in ALH84001.

However, the news had little effect on the course of world events. Wars were still fought. Politicians and celebrities went on making their accustomed headlines. Leading Christians and atheists made predictable responses to the ALH84001 story. So did others. The ecologist Colin Tudge said, "What is truly urgent is to study - and to preserve - the life that we already have on Earth, much of which is likely to disappear during the next few decades". This was a valid and laudable sentiment, but hardly relevant. Rather amusingly, the *Financial Times* announced "Discovery sparks hopes of fresh research and more funds", making their priorities clear. Indeed, several commentators drew attention to the timing of the ALH84001 revelation; it was announced while NASA's bid for further space exploration funding was being considered. Supporting NASA, the then President of the USA, Bill Clinton, declared that the discovery was "another vindication of America's space programme"; a curious assertion, since the meteorite had been discovered in Antarctica.

It is often said that contact with an extraterrestrial intelligence would be a momentous event. But would it? On the basis of the ALH84001 precedent, could we expect it to make much difference to us? Or would the world's response be "Very exciting, great story, but for all of us mortals it's business as usual"?

---

[38] UFO sightings occurred in flurries during the 1890s and the 1920s and then in 1947, 1952, 1957 and 1966. They have been interpreted as signifying visits to Earth (why?) by representatives of high-technology alien societies, usually housed on Mars or Venus, both of which are known to be lifeless. Despite the presumption that UFOs are associated with alien life, evolution was not discussed among UFOlogists until the later part of the 20th century. Interestingly, the frequency of UFO sightings over the course of history has been inversely proportional to the number of sightings of will-o'-the-wisps.

"Alien intelligence"[39]

The assumption "life-supporting planets elsewhere must entail alien intelligence" is not confined to UFO enthusiasts. We have already discussed the probability gulf between a potentially life-supporting planet and the emergence of multicellularity. *If* there are Earth-like planets, *if* they have characteristics compatible with life, *if* life chemically similar to ours ever began on them and *if* that life evolved with symbiosis-driven complexity increase, then *maybe* multicellular organisms evolved elsewhere in the universe. No step in this argument follows with great likelihood from the previous one. But the steps that follow are even less likely.

Let us take "intelligence" to mean an ability to respond to stimuli in novel (not pre-programmed) ways in order to solve real-world problems effectively. This behavioural definition makes no presumptions about internal processing of data by the organism. We shall use the more specific phrase "human intelligence" to signify a capacity (a) for detailed internal modelling of the perceived world and (b) to pass on information and material to future generations by non-biological means. This two-part definition is not behavioural. It embraces both human *mental powers* - our ability to relate what we see, hear and touch to stored memories via language and mental pictures, and *culture* - the passing on of tools, techniques, knowledge and social form to our children by teaching them, rather by genes.

We shall discuss this topic further in the remaining chapters, but two points are immediately obvious. First, human intelligence presupposes intelligence (e.g. culture is primarily a store of once-novel problem-solving techniques); but intelligence does not imply the inevitability of human intelligence. An organism might have novel problem-solving capability without anything akin to human mentation or culture. Second, when people talk about intelligence on alien worlds, what they mean, in our terminology, is *human* intelligence: language, culture and mental powers. "Human intelligence" in our sense of the phrase need not be confined to *Homo sapiens*. In some respects it might, for instance, be shared by other apes. But the phrase is convenient; it signifies the sort of "intelligence" that humans have, without implying that it is necessarily unique to our species.

---

[39] We use this word for convenience to mean any set of qualities that can distinguish "intelligent behaviour" from reflexive, pre-programmed or invariant responses to stimuli. It could be misleading; the set of qualities in question need not be the same in all animals, or always the same in the same individual or species. To suppose that "intelligence" denotes any thing-in-itself is an error. Here, we use the word as shorthand for a variable, open-ended set of descriptions of behaviour.

With all this in mind, let us consider the likelihood that a species with human intelligence will evolve, granted a planet with a thriving population of multicellular eukaryotes or their equivalent. On Earth it took roughly half the sun's expected life-span for human intelligence to emerge. How much more quickly could it have happened? It is hard to imagine human intelligence evolving in as little as - say - one tenth of the time it has taken ours, considering the number of evolutionary steps involved and their individual probabilities. So the statement "evolution of human intelligence takes a significant fraction of a star's life-span" could be generally true. This implies that although the universe might contain human intelligences with longer histories than ours, there cannot be any with *vastly* longer histories.

The fact that our sun is only one-third to one-half the age of the universe is not particularly relevant here. Life-bearing planets, indeed planets of any sort, could only come into being around second-generation stars like the sun; that is, stars made from the remains of stars that have already burned themselves out. The elements necessary for life - carbon, nitrogen, oxygen, phosphorus and heavier elements such as iron - are only made in stars nearing the ends of their lives. Therefore at least one generation of stars had to form, exist for a few thousand million years and finally explode before second generation stars with planetary discs could be formed and the universe could begin to bear life. The notion of a human intelligence that has survived for thousands of millions of years is therefore nonsensical, even if we ignore the fact that all species become extinct in much shorter times than that.

These arguments place general limits on the likelihood of human intelligence on other planets, but they do not rule it out. However, there are further arguments that cast serious doubt on the possibility. Suppose multicellular eukaryotes evolve on a planet several thousand million years before the star's intensifying radiation sterilises it – as happened on Earth. Would human intelligence necessarily, or probably, appear on that planet?

For organisms to behave "intelligently" in our sense they must be animal-like - able to move around to obtain food. How likely are animal-like creatures to evolve from primitive multicellular organisms? The probable answer is "not very". The time interval between the first Ediacara fauna and the first (Cambrian) animals on Earth was roughly the same as the time interval (around 600 million years) between the first animals and the present day. In other words, it did not happen quickly. This suggests that it was unlikely. However, suppose animal-like creatures *do* evolve. Will some sort of intelligence appear? This step is more probable: the ability to cope effectively with novel situations would surely be advantageous in any world of animals. But a major hurdle follows: given intelligence, would *human* intelligence evolve?

Gould and others would ask: why should it? Dinosaurs of little brain thrived for 140 million years. Mammals have "ruled the Earth" for less than half that time. Modern humans have existed for less than 1% of the mammalian era. Where was intelligence, let alone human intelligence, before the dinosaurs? All the dominant species of the Earth's past survived without anything akin to human intelligence. Even if we accept Kauffman's view of evolution rather than Gould's, it is very hard to see human intelligence as one of a choice of likely patterns. Ernst Mayr pointed out that during animal evolution, vision of some kind has evolved no fewer than twenty separate times, flight four separate times and human intelligence once. Vision has great survival value for animals; flight has considerable survival value for some animals - hence the instances of "convergent evolution". But what is the survival value of human intelligence (as opposed to intelligence in general)? It cannot be very great or it would have evolved more than once. Human intelligence seems to be a freak phenomenon, very unlikely to be repeated in any alien evolutionary system.

In summary: given multicellular organisms, animals seem unlikely to evolve. If animals do evolve, then some form of intelligence is quite likely. But given intelligent animals, human intelligence is extremely *un*likely. And as we reasoned earlier, the probability that multicellular organisms exist anywhere else in the universe is not very high to begin with. Reasoning from the evidence available to us, therefore, we are forced to conclude that human intelligence is an extraordinary and probably unique freak of nature, so there is almost certainly no human intelligence anywhere in the universe except on our own planet.

Advanced technologies
Suppose this conclusion is wrong. Suppose human intelligence *has* evolved on at least one other planet in the universe. Will it have produced a technologically advanced civilisation whose activities generate radio wave emissions? This inference is often supposed inevitable, but in truth it is far-fetched. Indeed, it is almost certainly false. The technologically advanced civilisation of our modern world is a consequence of the socio-economic system that emerged in western Europe after the 16th century, and particularly after the factory-based industrialisation of the 19th century. No comparable development had occurred during the previous hundred millenia or so of our species history, or in the previous 5,000 years of civilisation. It was a product of particular cultural conditions, not of biology, still less of physics. It was in no way predetermined or predictable. It was a product of very specific historical circumstances in just one culture. So it is

astronomically unlikely to have been replicated in any other species with human intelligence.

An obvious retort to the conjecture "A species with human intelligence will become technologically advanced" is to ask "Why should it?" This is simply unanswerable.

The Drake equation revisited

If these conclusions are correct, then the SETI project (Search for Extra-Terrestrial Intelligence) and the funding and attention given to it are absurd. SETI is based on serious misconceptions about the likelihood of alien intelligence and about our chances of communicating with it. In the light of our arguments in this chapter, let us reconsider the factors in the Drake equation. So far as R^* (the rate of formation of stars compatible with the development of life), f_p (the fraction of those stars that have planetary systems) and n_e (the number of planets orbiting each of these stars that has a life-supporting environment) are concerned, recent investigations have provided some clues. R^* and f_p probably have fairly high numerical values, but n_e is probably low. It is difficult decide whether f_l (the fraction of these planets on which life actually appears) is a reasonably high number or close to zero. It depends on the likelihood of the origin of life, about which opposing views can be held, as we have seen. However, we can be sure that f_i (the fraction of life-bearing planets on which intelligent life evolves) is vanishingly small, and f_c (the fraction of civilisations that develop an advanced technology emitting detectable radio signals) is effectively zero. The factor L is accordingly irrelevant. Serious omissions from the Drake equation include the fraction of planets developing intelligent life on which *human* intelligence evolves (again, this is almost certainly close to zero), and the components of f_i, which include the probabilities of developing eukaryotes, multicellularity and animal-like forms – collectively very low.

No doubt this will prove an unpopular conclusion, so let us suppose that we are wrong. Suppose, despite all reason to the contrary, we discovered an extraterrestrial intelligence with advanced technology and found a way of signalling to it. News that we were not alone in the universe might be reassuring, depressing or uninteresting, depending on our point of view (compare the various responses to the ALH84001 story). But what could we gain from such a discovery? The notion that we could usefully exchange ideas with hypothetical intelligent aliens is ridiculous. Consider: if we had a time machine that enabled us to exchange ideas with our own forebears three or four centuries ago, what could they gain from us or us from them? How could we possibly convey (for instance) the ideas of quantum physics, modern evolutionary theory, molecular biology, modern art, high-technology warfare, commercial pop music or the internet to Newton and his contemporaries? And at what level could we hope to grasp, or to respect, *their* way of thinking about the world? The comprehension barrier on both sides would be too great. How incomparably more difficult would it be to exchange ideas with an entirely alien culture from an entirely alien world? It is wholly implausible.

So much, alas, for the most enduring and entertaining theme in our science fiction literature.

Chapter 16

INTELLIGENT BEHAVIOUR AND BRAINS
The biological meaning of "intelligence"

In chapter 15 we introduced the notion of "intelligent behaviour". We describe behaviour as "intelligent" if it is capable of being is flexible and novel, allowing the animal to respond successfully even when it receives inadequate stimulus information. What kinds of animals have this quality?

- Intelligent behaviour requires a range of sensory input channels, some or all of which must have high capacity. The more ways the animal has of sensing its surroundings – in other words, the greater the range of stimuli it can perceive - the greater its capacity for behavioural novelty.
- The more ways in which the animal can respond to changes in its surroundings, the greater its capacity for flexible behaviour. So intelligent behaviour also requires a wide range of outputs.
- If the animal's behavioural outputs and sensory inputs are varied and of high capacity, then it is a very complicated task to select, sort and integrate the flow of information involved. Therefore, intelligent behaviour requires a *large brain* that can sort, integrate and correlate vast amounts of information quickly.
- Brain size alone is not sufficient. The more routes there are between stimulus and output, and the more indirect and cross-connected these routes become, the greater the flexibility and novelty of behaviour. Therefore, the brain of an animal that can behave intelligently has enormous numbers of indirect, cross-connected routes between inputs and outputs.

For an animal to make full use of a large brain with numerous cross-connections between multiple high-capacity input and output channels, there are at least two other requirements:-

- It must be able to store memories of past situations and behaviours, so the present situation can be compared with previous experience and the best course of action selected or devised.

- It must be capable of learning; i.e. distinguishing responses that are appropriate in a given situation from responses that are not. Observation of adult behaviour is important for learning by the young; the young remember and emulate what their elders do. Intelligent animals therefore tend to be social and to protect and "instruct" their young.

An animal's behaviour need not be "intelligent" just because the animal has a brain. The main functions of the brain are to control the animal's internal physiological activities and its responses to environmental stimuli. The second function is relevant to behaviour (intelligent or otherwise). To fulfil this function, the brain has to integrate all currently relevant sensory information. It must use this integrated information to create an internal representation or "model" of the environment, and of the animal's body in relation to that environment. Then it must direct the body's responses in accordance with this model.

Overall, animals behave in ways that ensure their survival and reproduction. They do this by responding moment by moment to changes in the environment and to signals from their bodies (for instance, perceived danger is avoided; food is sought when an animal is hungry). All behaviour, not least intelligent behaviour, is therefore targeted or goal-seeking. When an animal behaves "intelligently", memory and learning are crucially involved and most behaviour is generated internally, in the brain. Intelligent behaviour is *modulated* by stimuli from the environment, but does not usually *arise directly* from such stimuli. It appears rational, or creative, a matter of choice rather than reflex or instinct. Perception is an *active* process, seeking and selecting sensory stimuli in accordance with the animal's needs.

Primate (especially human) brains are tremendously complicated, but they are made of cells like any other organ; and each cell conforms to the "living state" model that we summarised in chapter 10. Cells in the brain exchange information with one another, as in the rest of the body. The cells that carry sensory information to the brain, process it and cause responses are nerve cells or *neurones*. Circuits of neurones relate sensory input to response (output) and do the work of learning and remembering. To understand "intelligent behaviour", therefore, the first step is to explore how neurones work. The second step is to explore how the junctions between neurones, *synapses*, perform their role.

How neurones work
A neurone is a terminally differentiated and highly specialised cell. Its appearance is unmistakable. Thin branching projections of various lengths grow like a dense copse of trees from the cell body. (The cell body is where the nucleus, mitochondria and other customary structural components are housed.) These projections are called *dendrites*. Their job is to pick up chemical or electrical signals, usually from other neurones. The signals picked up by the dendrites change the electrical potential in the cell body. We shall explain how this happens shortly.

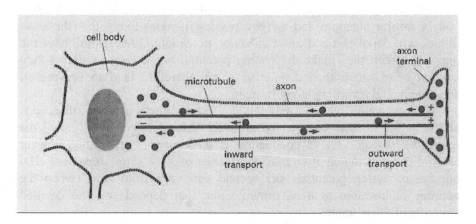

Fig. 16-1: a neuron, showing cell body, dendrites, axon and axon terminus. (See also chapter 5.)

One projection, the *axon*, has the opposite job. It conveys electrical impulses outwards *from* the neurone's cell body, not into it as the dendrites do. The axon arises at a small swelling known as the *axon hillock*. It ends in a *terminal arborisation*, rather as a big river might end in a delta. The terminal arborisation might comprise thousands of branches. Each branch ends in a *terminus*. The termini might connect with an *effector,* such as a muscle cell or a secretory cell in a gland; thus, an impulse travelling along the axon could make the animal twitch a limb or roll up into a ball, or secrete a noxious chemical. Alternatively, the axon termini might "connect" via synapses with the dendrites of other neurones.

In most cells, including neurones, the inside is electrically negative compared to the outside. The potential difference is about 0.05 – 0.1 volts[40]. In neurones, this difference is called the *resting potential*. Signals picked up by the dendrites alter the permeability of the membrane to certain ions, changing the resting potential locally. All these local changes are transmitted to the cell body, which adds them together. If the net effect is to decrease the cell body's resting potential sufficiently, then the axon hillock membrane becomes more permeable to ions, and for a few thousandths of a second the inside becomes positive relative to the outside. We say that the axon hillock membrane has been *depolarised*. A current immediately flows between this depolarised region and neighbouring parts of the axon, causing the latter to depolarise in turn. This effect is progressive; a wave of depolarisation travels along the axon. This wave of depolarisation is called an *action potential*.

Some stimuli tend to depolarise a dendrite membrane, lowering the cell body's resting potential and thereby making it more likely that the axon hillock will depolarise and cause an action potential. Other stimuli have the opposite effect: they make the resting potential bigger, not smaller – they *hyperpolarise* rather than depolarise – and therefore tend to prevent an action potential travelling along its axon.

After an action potential has left the axon hillock, a few milliseconds elapse before another one can be initiated. This brief delay is called the *refractory period*. Nerve conduction is therefore not a steadily flowing electric current; rather, it comprises a series of very short impulses. The number of action potentials per second can vary from zero (when the neurone is inactive) to a maximum value that depends on the detailed structure of the neurone.

Synapses
A tiny gap, just a few nanometres across, separates the terminus of each axon branch from the next neurone. This gap is a *synapse*. Neurones "talk" to each other across synapses. When an action potential reaches the axon terminus it causes the release of a small package of a special chemical substance, a *neurotransmitter*. The neurotransmitter crosses the synapse and binds to receptors on the dendrites or cell body of the next (postsynaptic)

---

[40] This effect has quite a simple explanation, which was found by Donnan early in the 20th century. The main component of the membrane potential is the "Donnan potential" – a simple physico-chemical phenomenon. Nothing specifically biological (or magical) is involved.

neurone. When a receptor is occupied, the local membrane is either slightly depolarised or slightly hyperpolarised, so the electrical potential in the postsynapic cell body is changed, making it either more or less likely that an action potential will travel along the axon.

Picture a neurone (N) with two dendrites, each of which forms a synapse with an axon terminus from a different neurone. If one of these presynaptic neurones hyperpolarises its dendrite and the other depolarises its dendrite, then the response of neurone N will depend on the sum of the two inputs. The faster the action potentials in the depolarising axon are compared to those in the hyperpolarising one, the more likely N is to be activated, i.e. to transmit action potentials along its own axon. Thus, the rate at which a neurone "fires" depends on the sum of its current inputs. In a real neurone in a mammal's brain there might be ten thousand or more inputs rather than two; but the principle is the same.

How are the neurotransmitter packages[41] assembled in the axon termini? We mentioned this briefly in chapter 5. Neurotransmitters are made in the cell body and are packaged inside small membrane-bound vesicles. These vesicles are taken to the ends of the axon by a motor-driven process using fibres of the cytoskeleton, which run all the way along the axon like railway lines. An action potential makes some of these vesicles fuse with the axon terminus membrane, releasing their contents into the synapse. The empty vesicle is carried back to the cell body along the cytoskeletal fibres, to be re-loaded with fresh neurotransmitter and returned to the terminus for further use.

Circuits of neurones

Suppose a stimulus is detected by touch, vision or some other sense. Suppose the animal needs to respond rapidly by moving towards the stimulus source (if it means food) or away from it (if it means danger). This might be achieved through a *reflex arc*. The sensory organ stimulates neurone A. Action potentials in neurone A activate neurone B, which in turn activates neurone C. Neurone C brings about the required response.

---

[41] How many different neurotransmitters are there in mammalian brains? There seem to be several hundred; more are discovered every year. Some are quite simple molecules, often derivatives of amino acids. Others are peptides (fragments of proteins), often quite large ones. Different neurotransmitters have different effects on postsynaptic cells; they might activate or inhibit; the effects might be transient or longer term. All neurones specialise in the neurotransmitters they make, but most neurones make more than one. Therefore, a single action potential can have different effects on different postsynaptic neurones.

Dedicated pre-programmed circuits of this sort are found in all animals with organised nervous systems. Reflexes are staple parts of all animal behaviour. Reflex behaviour is not "intelligent" because the stimulus is *directly* and *inflexibly* linked to the response. There is no variability, no choice, no requirement for learning or memory, and the response is caused by the stimulus, not internally in the brain. Nevertheless it is possible to intervene in these dedicated circuits. Suppose neurone B is inhibited by another neurone, X. When X is activated it blocks the A-B-C circuit and diminishes or even eliminates the reflex. Thus, reflexes can be overridden. Alternatively, if X weakly stimulates B, it will *potentiate* the reflex, and this can lead to an elementary form of learning. Suppose X is stimulated just before A is stimulated. If this sequence of stimuli is repeated often enough, it can change the circuit permanently. In time, the stimulus through X will elicit the response as effectively as the stimulus through A. This is "conditioned learning", a process made famous by Pavlov and his dogs[42].

More subtly, if one neuronal circuit is activated immediately before another, and if a neurone in circuit I interacts with a neurone in circuit II, then *learning by association* can result. After the same activation sequence has been repeated many times, the input to circuit I will evoke the output of circuit II. Associative learning was first postulated by Donald Hebb in the middle of the 20th century. It might partly account for an intelligent animal's ability to predict events. There is evidence that Hebbian associative learning happens in a wide variety of animal species.

Intelligent behaviour, which originates in the brain rather than from external stimuli, must involve far more complicated processes than conditioned reflexes and associative learning; but it too must require interacting circuits of neurones. The "models in the brain" that we mentioned in the first section of this chapter must be particular sequences of circuits through which information is processed and behavioural outcomes are implemented. Information seems to "coded" in these circuits by – among other things – the frequencies of the action potentials and the actual location of each circuit in relation to others; but here we reach the present limits of neurobiological knowledge

---

[42] Ivan Pavlov pioneered the strategy of reducing behaviour (and mental processes) to physiology. He discovered that if a bell was always rung immediately before his dogs were fed, then after a time the dogs began to salivate in response to the sound of the bell, irrespective of whether food was offered. Conditioned learning is critically dependent on the temporal order of the stimuli. Pavlov would not have obtained interesting results if he had rung the bell immediately *after* feeding the dogs.

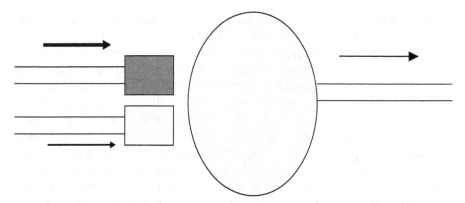

Fig. 16-2: associative learning. A schematic outline of Hebb's hypothesis. Hebb suggested that memories might be formed by the mechanism outlined in this simplified diagram. Two incoming neurons (on the left of the picture) synapse on to a third. The cell body or dendrites of the third neuron are illustrated by the large ellipse, and the axon of the third neuron extends to the right of the picture. The input from the upper incoming neuron is strong enough on its own to make the third neuron fire, but the input from the lower incoming neuron is too weak to elicit a response. However, if both incoming neurons fire simultaneously a sufficient number of times, biochemical changes occur that strengthen the lower synapse. After these changes, an input from the lower incoming neuron suffices to make the third neuron fire, even when the upper incoming neuron is 'silent.

Embryonic neurones are called *neurites*. They grow towards specific targets along fibres of the extracellular matrix (the meshwork to which many of the body's cells are anchored). This growth lengthens them dramatically; that is how axons are formed. Both the target and the extracellular matrix are essential for neurite growth and therefore for nervous system development.

Chemical signals secreted by the target determine the direction of growth and also sustain the survival and maturation of the neurites. Each neurite has a set of receptors to ensure that it heads towards the appropriate target. It is possible to interfere experimentally with this process. Eliminating the target, introducing an artificial chemical gradient or altering the receptors will send the neurite in the wrong direction (or kill it). Such experiments have improved our understanding of embryonic brain development.

The extracellular matrix is necessary for an organised system of neurones to develop as the embryo matures. It is also necessary for establishing synapses, and for signals from the target to be correctly "interpreted" by a neurite. However, it is only required during the embryo stage. It is absent from mature brains. Like scaffolding, it is dismantled when the building is complete. Therefore, the growth of new axon branches and new postsynaptic dendrites in the mature brain – i.e. the formation of new synapses - takes place without any "support system".

Brain development is just one aspect of embryo development, a topic that we touched on in chapters 8 and 9. Embryo development involves the sequential expression and suppression of various groups of genes, a process roughly analogous to a chord progression in music. Each new pattern of gene expression is associated with a new internal state of the cell and responsiveness to a new set of stimuli. Also, the cell's ability to *send* messages to other cells is changed. Within each cell, the three-way reciprocity of internal state, responsiveness to signals from other cells and gene expression pattern causes a programmed progression of changes. This programmed progression is initiated by the expression of just one or two genes, known as *immediate-early genes*. During the execution of the developmental programme, hosts of other genes are expressed and suppressed in every cell.

The key genes in brain development are equally crucial for the development of other organs. In the favourite species of geneticists, the fruit-fly *Drosophila*, gene defects that alter behaviour, learning and memory also alter muscle activity, female fertility and other functions quite remote from the brain. It is therefore absurd to speak of genes "for" behaviour, memory or learning, and particularly absurd to speak of genes "for" a particular *type* of behaviour. Rather, we have genes that play key roles in embryo development generally. A defect in one of these genes will lead to abnormal development, including abnormal brain development. This will result in deviations from normal behaviour, along with other anatomical or physiological anomalies[43].

Synapses, learning and memory

Animal cells exchange signals with one another throughout life, not just during development. These exchanges are necessary if the organism is to function as an integrated unit and survive. Neurones are no exceptions, though their method of communicating - by chemical signals at synapses - is specialised. Any cell might alter its internal state and gene expression pattern in response to a signal from another cell. Again, neurones are no exceptions; but again, their signal receptors are confined to membranes bordering the synapse.

[43] This point is widely misunderstood. One example will serve for illustration. Some years ago, newspaper headlines heralded the discovery of a "gene for the nurturing instinct" in mice. What had *really* been discovered was that elimination of a particular immediate-early gene known as *fos-B* produced poor mothers. These genetically damaged mice did not groom their pups normally, or show any urgency in recovering them when they wandered. Mice lacking *fos-B* develop abnormally in a brain region known as the preoptic area, which in normal mice becomes active in many stimulus situations - including presentation of pups. So the *fos-B* deficient mice had impaired development in a significant part of the brain, and in consequence had a number of behavioural abnormalities. *One* of these was poor maternal behaviour. A phrase such as "gene for the nurturing instinct" betokens a misunderstanding of biology.

After the embryonic brain has developed, the responsiveness of neurones to signals becomes important in learning and memory. Two main types of process seem to account for at least some aspects of learning and memory: structural remodelling of the synapse, i.e. the formation of new axonal branches and perhaps the removal of old ones; and functional changes in synaptic strength. Both these mechanisms depend on transmitter chemicals that alter the postsynaptic cell's internal state and its pattern of gene expression. These signalling chemicals usually occupy their receptors for longer periods than neurotransmitters do.

Synaptic remodelling works roughly as follows. Suppose a chemical signal released from the presynaptic neurone causes one or more long-term changes the postsynaptic cell (1 in Fig. 16-3). This cell responds by secreting a factor that enters the presynaptic neurone (2), where it is carried back to the cell body along the cytoskeletal transport system (3). Here, it alters the gene expression pattern of the presynaptic cell. One consequence might be a redistribution of membrane material at the axon termini (4) and growth of new axon branches (5), perhaps leading to the formation of new synapses with other neurones. This kind of remodelling has been described in large animal brains. The implications for learning are considerable. Although mature neurones cannot divide or be replaced, they can continue to grow and to make new branches and new connections. The phrase "life-long learning" makes biological sense.

The alternative to remodelling is to change the strength of an existing synapse. Long-term potentiation (LTP) and long-term depression (LTD) seem to account for some aspects of memory. LTP depends on a chemical, secreted from the axon terminus, which only occupies its postsynaptic receptors when it is released in sufficient quantity. This happens when the axon transmits rapid successions of action potentials at frequent intervals. Occupation of the receptor causes the postsynaptic cell to take in calcium, which changes certain signalling pathways, activating a succession of genes that, ultimately, make the postsynaptic membrane permanently more likely to depolarise when it is stimulated.

It is easy to imagine that LTP provides a mechanism for Hebbian associative learning and for more complicated learning processes. If two stimulus-response circuits interact, then LTP of the synapses that link them makes it easier to transmit an impulse from one circuit to the other.

Fig. 16.3 Synaptic remodelling: an outline

An action potential travelling along the axon of the presynaptic neuron releases a chemical signal (1), which binds to a receptor on the postsynaptic membrane (2), allowing calcium ions to enter the cell. The calcium ions travel to the nucleus in the cell body and activate certain genes (3). When these genes are expressed, the proteins made travel to the postsynaptic membrane, where they are incorporated (4). This changes the responsiveness of the postsynaptic membrane to certain neurotransmitters, so the behaviour of the synapse is permanently altered. At the same time (not shown in the diagram), similar changes take place in the pre-synaptic membrane. A 'retrograde signal' released from the postsynaptic cell activates changes in the presynaptic membrane, leading to changes in gene expression and alterations in the presynaptic cell.

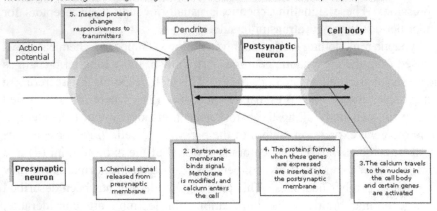

Fig. 16-3: synaptic remodelling, the possible cellular basis of learning and memory.

However, the best available evidence suggests that LTP is relevant only to *long-term* memory storage, which is established only an hour or two after the learning event has taken place. It is not relevant to immediate and short-term memory. As an explanation for associative learning, therefore, it is not entirely adequate.

LTD, which makes the postsynaptic membrane becomes less liable to depolarisation, involves an analogous mechanism. Whereas LTP is a response to rapid repeated use of the circuit, LTD results when the use is slow and prolonged. Thus, LTD might enable neuronal circuits to adapt to continuous, and therefore uninteresting, stimuli. It stops the brain paying attention to them. Without such adaptation, a brain could not function effectively. It could not focus on significant changes in the animal's environment.

What sorts of animals can behave intelligently?

Our broad definition of "intelligent behaviour" applies to many types of animals. It is not confined to warm-blooded vertebrates. Octopus behaviour is as "intelligent" as the behaviours of most mammals, though brain organisation is very different in mammal and octopus and the responses to stimuli are quite distinct. In some instances, intelligent behaviour appears to reach across generations. The mechanisms involved are mysterious. A honey-bee swarm returning from migration readily locates its original nest site, though no individual in the swarm has been there before, the original inhabitants having died. Do particular swarms leave distinctive chemical markers that are durable enough for their returning descendants to recognise? This hardly seems plausible, but it is difficult to see any other explanation. Salmon returning to the rivers in which their parents bred take refuge during the journey behind the same rocks that their forebears used, although alternative, equally adequate, shelter is available. These remarkable phenomena remind us that much about animal life continues to defy ready explanation.

Interesting as these reflections are, we shall focus on mammalian brains for the rest of this chapter. This will serve as a prelude to discussing "human intelligence" in chapter 17. Mammalian brains consist of three main parts: *forebrain, midbrain* and *hindbrain*. Very broadly, the hindbrain controls basic physiological processes - heart rate, breathing, body temperature, eating and drinking, sleeping and waking. The midbrain co-ordinates sensory information from body and environment and initiates appropriate responses. Parts of the midbrain are associated with emotions and with aspects of memory storage. The forebrain is the area most concerned with intelligent behaviour. In primates in particular, it is dominated by the *cerebral cortex*. The cerebral cortex is fairly small in mice but covers the entire brain in gorillas, chimpanzees and humans. A mouse deprived of its cerebral cortex behaves quite like a normal mouse, but a human deprived of his or her cerebral cortex is a vegetable. It is tempting to infer that the cerebral cortex is the root of intelligent behaviour. This seems to be the case for mammals, but it is unwise to generalise too much. Birds, which have no cerebral cortex at all, are often capable of "intelligent behaviour" in our sense of the phrase. But because we are now focusing on mammals, the cerebral cortex deserves particular attention.

The cerebral cortex
This structure consists of four main segments or lobes: *frontal, temporal, parietal* and *occipital*. Each of the four has various complex functions. The occipital lobe, at the back of the brain, does most of the processing of visual information. The parietal, across the top of the brain, co-ordinates the body image (from the sense of touch and the internal receptors that detect information about balance and posture). It also oversees body movements. The tasks of the temporal lobes, one on each side of the brain, include

processing auditory information. The frontal lobe is concerned with - among many other things - cognition and abstract thought. As a very rough rule of thumb, the front half of the cortex controls outputs such as muscle movements and the back half is concerned with processing sensory inputs; but this division is far from absolute.

Sensory inputs come from two main sources: the environment, and the animal's body. In mammals, the environment is detected by seeing, hearing, touching, tasting and smelling; the relative importance of these five main sensory modalities varies from species to species. In humans, some 85% of the information about the environment processed by the brain is visual; the visual areas occupy a large part of the cortex, mainly the occipital lobe. Information from the body includes sensations of position and movement in head and limbs, sensations of gravity and acceleration, and basic physiological operations such as heart function, respiration and nutritional status. The brain – and in particular the cerebral cortex - has the task of relating all the information from these two main sources moment by moment and executing the appropriate behavioural outputs.

Throughout the cortex, neurones are organised in cylindrical "modules" with their axes perpendicular to the surface of the brain. There are about 150 neurones in each module (rather more in the visual areas of the occipital lobe), arranged vertically in six layers, with layers of horizontal neurones between. Each neurone can make connections with hundreds, thousands or tens of thousands of others, both within the cortex and in the rest of the brain. Their cell bodies bristle with dendrites and their axons have massive terminal arborisations. Because the number of connections among cerebral neurones is so astronomical, it is easy to forget that the whole structure of the cortex is organised in simple repeating patterns. But much of our knowledge of cortical function has come from studying neurones in particular cylindrical modules.

The cerebral cortex is by no means independent of the rest of the brain. For example, it is continuously activated by a "motivational" system in the centre of the brain, the *basal ganglia*, and in particular the *corpus striatum*. These structures are activated by centres in the hindbrain and cerebellum that process information from the body; they receive inputs from the cortex; and they are crucially important in controlling and co-ordinating movements.

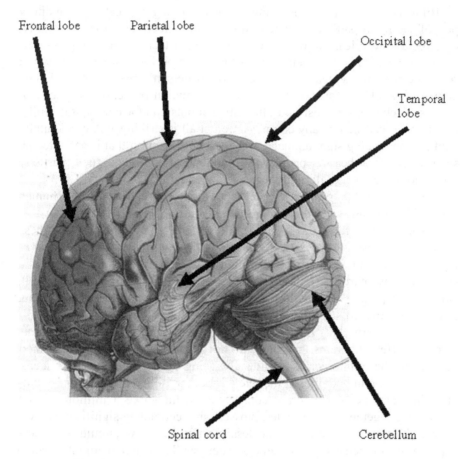

Frontal lobe

Parietal lobe

Occipital lobe

Temporal lobe

Spinal cord

Cerebellum

Fig. 16-4: the human cerebral cortex and cerebellum.

The *hippocampus*, located below the temporal lobes, processes visual information before it reaches the visual cortex and is also of primary importance in storing memory, especially spatial memory. All parts of the brain, including the cortex, are interconnected.

Plasticity of brain function
There has been a long-running debate about whether the brain behaves "holistically" (all parts are responsible for all functions) or whether each little piece of the brain has its own particular job. In a sense, both views seem to be correct. Each part of the cerebral cortex has a distinct function, but most of these parts are multiply connected to each other and to the rest of the brain. As a result, if some areas in the cerebral cortex are destroyed, their functions are lost; but other areas are more "flexible" – if they are damaged, the rest of the cortex can compensate.

Blind cats are better at locating sounds than sighted cats: the auditory part of the brain partially compensates for the defective vision. Similarly, blind people who learn to read braille have better tactile processing systems in their brains; so have jewellers and others who perform fine manual work. A braille message activates a part of a blind person's brain that in sighted people responds to visual stimuli. (The information processing capacity of the visual channels is vast and alternative inputs cannot compensate fully, but the fact that there is any compensation at all is striking.) A congenitally deaf person reading sign language uses a part of the brain that is activated by speech sounds in hearing people. Mammalian brains retain such plasticity throughout life, though it might decrease with ageing.

However, if the *primary* visual and auditory regions in the human cerebral cortex are lost (the regions where information from the retina or inner ear is first recorded and sorted), the sufferer is blind or deaf and these losses cannot be compensated. Plasticity is confined to the *secondary* regions, where the perceived information is interpreted. Similarly: if a stroke destroys part of the motor cortex then the patient might recover well. But if other parts of the brain, for example the basal ganglia or cerebellum, have been damaged, then there is hardly any recovery.

Like most terminally differentiated cells, mature neurones cannot divide. In most vertebrate brains, neurones are not replaced when they die, so their number in the brain decreases with advancing age. After their late teens, humans lose brain neurones at the rate of about a million a day. This sounds alarming, but the human brain contains something like a million million neurones altogether. Nevertheless, losses can accumulate significantly over a long life-time. If the brain had less plasticity, if new circuits were less capable of compensating for damage, then senile dementia might develop much earlier.

Brains and computers

Mammalian brains, particularly human brains, are the most complicated objects known in the universe. Every generation compares the brain to the most complicated piece of technology so far invented. In the 17th century, Leibnitz compared it to a water-mill. At the end of the 19th century, Freud compared it to a hydraulic system. In the 1930s it was compared to a telephone exchange, in the 1960s to a digital computer. Most recently it has been compared to a neural network or parallel-processing system, a development in computer technology partly inspired by (but not necessarily intended to simulate) a living brain. How useful is this analogy?

Neural networks, like brains, have no central processing units of the kind found in digital computers. Different areas of the network, as in the brain, engage in democratic dialogue. And like brains, neural networks can detect signals, recognise patterns, interpolate data, make predictions, guide movement on the basis of visual information, and even synthesise speech. But they are not really like brains. A one-year-old child's visual processing capacity far exceeds that of any computer. Different areas of the human brain detect and interpret form, motion and colour in a *single* visual stimulus. The fine division of labour here, and the vast number of neurones involved, distinguishes the brain from a neural network. Also, a three-year-old child's language production, which again involves very fine division of function among closely related brain areas, is *qualitatively* different from anything that a machine can do. In particular, neural networks do not remember or learn in anything like the way brains do.

This is not to say that computer models or analogies of the brain are useless - or that there is anything wrong with neural network systems. In the final chapter we shall explore a computer metaphor that recalls our general model of the living state. However, brains have to be studied as objects in their own right. Brains are brains. They are unlike any piece of technology we have or are ever likely to have. They are not soggy computers, any more than they are soggy water-mills.

Chapter 17

HUMAN EVOLUTION
Human intelligence and the question of human uniqueness

Intelligent behaviour is a device for ensuring the survival of complex, mobile organisms. All primates show intelligent behaviour. The question we now face is whether *human* intelligence is qualitatively different. Is our species unique in any non-trivial way? We shall approach this question by looking first at human evolution, then at the distinctive features of human brain function, and finally – briefly – at the notion of "mind" as a biological entity.

Human evolution: an outline
Details of human evolution are controversial but some points are generally

- The evolutionary lines that led to modern chimpanzees and modern humans probably diverged between five and seven million years ago.
- The common ancestor of humans and chimpanzees behaved *intelligently* in the sense explored in chapter 16 (learning, predicting, solving problems correctly on the basis of incomplete data, exhibiting flexibility and novelty of behaviour, and so on).
- Although many of the connections between one stage of human evolution and the next are uncertain, a plausible chronology of hominid types can be constructed (see the diagram).
- The most dramatic feature of human evolution was a rapid increase in brain size relative to body size. The evidence for this comes from skull fragments, which enable us to calculate cranial capacity and hence brain size. Modern chimpanzees have somewhat smaller bodies than modern humans, but very much smaller brains. An adult chimpanzee's brain measures 450 cc; an adult human's

measures 1300-1400 cc. In other words, the size of the adult human
brain has tripled during a mere six million years or so of evolution.

- Before cranial capacity (brain size) increased significantly, hominids
  became bipedal. Most people[44] agree that this development was
  crucial. Walking on two legs liberated the hands for tool use. Tool
  use preceded and may have been instrumental in "causing" the
  expansion of the brain.

- In the most recent evolutionary steps, from *Homo habilis* to *H.
  erectus* and finally *H. sapiens*, much of the increase of brain size
  probably involved the frontal lobe[45].

- From the earliest hominids onwards, our ancestors seem to have
  been very highly social. The genus *Homo* has probably produced
  the most social mammals ever. Individuals have always depended
  for their survival on close co-operation within groups.

All stages of human evolution except possibly the most recent have taken
place in Africa. The very oldest Australopithecines[46] – presumed to be our
earliest ancestors after division from the chimpanzee line - have been found
in Ethiopia and South Africa. They or similar species probably occupied the
part of the continent between these two areas. Opinion is divided about
whether *H. sapiens* (1) originated in Africa like all its forebears and then
migrated into Asia and Europe, or (2) evolved from *H erectus* after the latter
had migrated to other parts of the world. The first alternative, the "out-of-
Africa hypothesis", has powerful support. A very early *H. sapiens* grave
(about 55,000 years old) has been found in Upper Egypt, and no comparable
finds of the same age have been made outside Africa. This implies that
sapiens originated in Africa. Also, some *H. erectus* remains in Java date
from a time (30-50,000 years ago) after *H. sapiens* had reached that part of
Asia, so the two "species" presumably co-existed for a time. This finding is
difficult to reconcile with the second alternative, the "multiple origins
hypothesis". On the other hand, some ancient Australian art appears to be

---

[44] One of the first people to propose this idea was Frederick Engels. Long before anything
was known about the ancestry of modern humans, Engels argued (for political reasons)
that technology must precede knowledge; labour is the antecedent of thought; the work of
the hand leads the work of the mind. So far as human evolution is concerned, he appears
to have been correct.

[45] It is customary to regard *H. habilis, erectus* and *sapiens* as three different species.
However, it is hard to be sure that they could never have interbred.

[46] The prefix "Australo" indicates "south" rather than Antipodean. The remains of these
earliest of human ancestors, up to 3-4 million years old, were first found in southern
Africa.

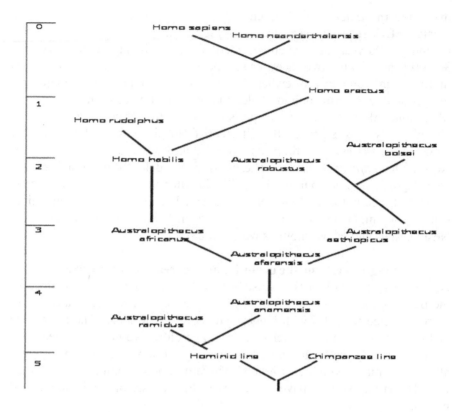

Fig. 17-1: a possible outline of the evolution of *Homo sapiens*. The vertical axis is a time scale indicating millions of years before the present.

more than 50,000 years old. These creations were presumably the work of *H. sapiens* not *H. erectus* (see below), and *sapiens* was unlikely to have reached Australia 50,000 years ago if the "out-of-Africa" hypothesis were correct. So there is some support for the "multiple origins" hypothesis. The issue is far from settled.

At various stages in human evolution two or more distinct races, or subspecies, or closely-related species of hominid must have met each other. When this happened they would either have ignored each other, mated with one another or killed one another. Probably they followed all three options at different times and in different places.

This applies not only to *sapiens* and *erectus*, but also to the more recent interaction between modern humans and Neanderthals (30-25,000 years ago). It is still not clear how the Neanderthals were connected to the rest of the hominid line of descent. They were heavier-boned and heavier-muscled than modern *sapiens* but their cranial capacity was the same or slightly greater.   Evidence suggests that *H. sapiens* entered what is now Europe around 35,000 years ago, when Neanderthals were already in residence.   By 30,000 years ago the two races, or subspecies, had met at various locations in Iberia and probably elsewhere.   Since the Neanderthals subsequently disappeared, consensus opinion holds that modern humans wiped them out. Mitochondrial DNA evidence supports this view: the mitochondrial DNA of Neanderthals was significantly different from that of modern humans. However, skeletons have been discovered that are unquestionably those of Neanderthal-*sapiens* hybrids.   These skeletons date from around 25,000 years ago, i.e. some four or five millenia after the two races first made contact.   Since mitochondrial DNA is inherited solely from the female line, some of us might possibly be descended from hybrids whose fathers were Neanderthals and whose mothers were *sapiens*.

The "driving forces" for the evolutionary increase in brain size
Evolutionary psychologists have suggested that the large human brain, like the peacock's tail, is the result of sexual selection.   Ancestral peahens were more attracted to mates with bigger and more ornate tails.   Therefore, size and ornateness of tail were selected for.   Therefore, we now have peacocks with ludicrously exuberant tails.   By analogy, hominid females were more attracted to mates with bigger brains.   Therefore, bigger brains were selected for.   Therefore, we now have humans with ludicrously big and complicated brains.

On the face of it, this seems plausible.   Sexual selection has been held to account for evolutionary "super-growth" in various animal features. However, sexual selection only favours exaggerated development of features in the *male*.   It is the peacock, not the peahen, that has the dazzling tail.   In humans, the ratio of brain size to body size is essentially the same in both sexes.   Moreover, even if brainier males *did* attract more mates in the evolutionary past, this would not explain why their ancestors started to become brainier in the first place.   The sexual selection hypothesis of human brain evolution might contain a grain of truth, but it cannot be the whole truth.

Let us consider an alternative "just so story".   It is generally agreed that bipedalism predated the increase in hominid cranial capacity.   What were the immediate advantages of bipedalism?   First and foremost, it freed the hands for tasks other than locomotion, such as using tools.   Some elementary use of *objets trouvés* as tools probably began very early in human evolution, or even before; other ape species, such as chimpanzees, can also use tools.   The individuals who were best at using tools were likeliest to survive and leave offspring, so the capacity for tool use was selected for.   Greater capacity for tool use implies better hand control, which could only have been achieved by increased development of brain areas concerned with hand movements and sensations. Therefore, natural selection favoured the growth of those parts of the cerebral cortex that process sensory information from the hands and control the fine movements of digits[47].

The better the control of hand movements became, the greater the possibilities for tool use. As a result, the range of tool uses increased. In time, hominid populations became more dependent on tools. This made it increasingly advantageous for the young to learn about tools: how to find them, how to use them and - later - how to make them. A brain that was good at learning from adults in the community, acquiring and developing skills and passing on those skills to others, became useful and then indispensable. Ultimately, teaching tool-using skills to the young became necessary for the survival of the next generation. Therefore, the winners of this part of the evolutionary race were hominids whose brains were best at learning and communicating. They were also best at distinguishing quickly and accurately among individuals within the community. It was clearly advantageous to be able to recognise adults who could teach essential skills, and adults who were best avoided.

By the time tool use had advanced from the incidental and occasional to the essential-for-survival, the hominid brain had (according to our scenario) become good at hand control, learning, communicating, and recognising individuals: in other words, at manual skills, language and facial recognition. The learning was predominantly *early* learning, so the period of childhood – i.e. dependence on parents and other adults - must have gradually grown longer as learning became more elaborate and the need for it greater.

Early learning forges new synaptic connections, increasing brain size. One of the most remarkable features of the modern human brain is its increase in volume between birth and adulthood. Both human and chimpanzee babies have roughly 350 cc brains at birth. An adult chimpanzee has a 450 cc brain, an increase of about 30% during maturation. But an adult human has quadrupled in size to 1200-1400 cc. This difference is almost entirely due to the formation of new axon branches and new synapses, the physical concomitants of early learning and environmental influence.

The brains of the earliest known Australopithecines were about the same size as those of chimpanzees. Even their hominid successors had cranial capacities of only 600-700 cc. At this stage in evolution, the part of the

---

[47] In the brains of modern humans, far more of the sensory part of the cerebral cortex is devoted to the hands than to any other part of the body except the genitals. The hands are similarly over-represented in the motor cortex. We propose that this "takeover" began in the very earliest hominids.

cerebral cortex that had grown most was presumably concerned with hand control. Not until *H. habilis* appeared was the 1000 cc barrier broken. By this stage, culture was probably elaborate enough for significant communication and facial recognition skills to have developed.

Of course, this is a much over-simplified picture. Many other factors must have been at work. Horrobin drew attention to the fat content of human brains (and human milk), which is distinguished by extremely high levels of certain unsaturated fatty acids. The richest sources of these fatty acids are bone marrow and small aquatic organisms. Significantly, early hominid communities lived near lakes or rivers and their diet seems to have included bone marrow, which is not a part of the diet of other apes. This idea, that some early hominids were semi-aquatic, is consistent with – among other features - the development of elaborate vocal communication, which demands considerable breath control and is found also in whales and dolphins. If there was a semi-aquatic phase during human evolution, many aspects of the development of the large human brain – including language capacity - can be explained.

Has the human brain stopped evolving?
The proposal outlined in the previous section presumes a dialogue between brain size and *culture*. But what we now mean by "culture" is incomparably more elaborate than anything experienced by our remote ancestors. Does the dialogue still continue?

As far as we know, *H. sapiens* has been the only extant species of *Homo* for at least 20,000 years. Cultural change during this period has been radical. Refinements in stone tools, and probably in social organisation among nomadic groups, led some 9-10,000 years ago to the first hints of a settled way of life and the beginnings of civilisation. Traces of einkorn wheat[48] from that time have been found in the Karacadag Mountains of south-eastern Turkey. Its original cultivation has been attributed to a single tribe. Settled agriculture led to increases of human group size and division of labour. In due course it led to cities, kings, scribes, craftsmen and professional soldiers, and to written language.

In the Euphrates valley, some way downstream from the Karacadag range, writing was allegedly invented by the Sumerians around 5,300 years ago. This momentous event might have been part of a steady cultural progression rather than a sudden novelty, as is often supposed. Oval stones with pictogram carvings dating from about 10,000 years ago have been found near the Euphrates in Syria. What these carvings denote is a mystery,

---

[48] Einkorn wheat was one of the "founder crops" of Neolithic agriculture. It still grows wild over much of the Middle East and the Southern Balkans.

but they imply the use of abstract symbols, which is the essence of writing. We might therefore infer that the first glimmerings of written culture coincided with the first glimmerings of settled life and civilisation. Thus, Sumerian writing could have been the culmination of a four or five millenium cultural progression in symbolic expression, not a radical innovation; just as the Sumerian cities were the culmination of a progressive development in settled community structure.

A written culture in a large, settled, differentiated, civilised community placed greater demands on learning and memory than life in a small, undifferentiated, preliterate nomadic tribe. Today, people from nomadic tribes can adjust to modern city life within a generation or two, showing that their brains have all the requisite capacity. The simplest inference is that civilised culture has evolved during the past 5000 years without any further significant changes in the human brain. However, there is probably no one alive today whose ancestors have, without exception, avoided all significant contact with civilisation. Perhaps many *Homo sapiens* 10,000 years ago could not have coped with the learning and memory demands of even rudimentary civilisation, let alone the modern city; but natural selection has eliminated them. Therefore, it is possible that the progress of *H. sapiens* towards civilisation has entailed continuing expansion of the brain.

Evolutionary psychologists reject this view. They assert that humans have not evolved since the Pleistocene. Therefore, modern people are endowed with "innate Pleistocene dispositions". This causes us to kill one another and practise infanticide, males to rape, and females to fall for rich men. This assessment purports to be Darwinian, but it depends heavily on an analogy between humans and scorpion flies, which seems tenuous. Its proponents account for inconsistencies between their hypothesis and observed fact by invoking "free will", a notion that Darwin rejected. The writings of evolutionary psychologists always seem to exude a "fall-of-mankind" pessimism, which may explain why they are fashionable. Why *should* the evolutionary dialogue between brain and culture have ceased? Evolutionary psychologists do not consider this question.

Unfortunately, the evidence on this point is not decisive. Adult human/hominid brain volume has increased by about 300% over a period of roughly six million years, which amounts on average to 1.5% every 30,000 years. The earliest *H. sapiens* whose cranial capacities can be reliably estimated are in the order of 30,000 years old; so all other things being equal, we would expect a 1.5% difference between their brain volume and ours. But 1.5% of 1400 cc is only 21 cc, which is less than the measurement error and less than the standard deviation in modern humans. Thus, the evidence does not support the claim that modern human brains are bigger than those of the earliest *sapiens*; but nor does it refute it.

On the other hand, we keep people alive today who in the past would have died in infancy because of genetic defects. For example, colour blindness has probably doubled in frequency since the earliest *sapiens*. Perhaps this "weakening of the gene pool" militates against further increase in complexity of our species and therefore against further increase in brain size. In any case, the limits of our mutation-correction machinery might have been reached in an organism as complicated as the modern human (see chapter 13); modern humans might be close to the theoretical limit of organism complexity. So there are plausible arguments that the human brain *will* stop increasing in size and complexity - but there is no compelling reason to suppose that the limit has been reached yet.

Distinctive features of human brain function
Human brains are bigger relative to body size, and much more complicated in terms of numbers of synaptic connections, than the brains of other mammals. However, being bigger and more multiply connected does not mean that they are better at everything than the brains of other species. For example, dogs process olfactory information far more efficiently and elaborately than humans. Salient examples of human capabilities are hand control, facial recognition and language; we discussed hand control and its evolutionary significance earlier.

An obvious prerequisite for facial recognition is vision. A good deal of the cerebral cortex, particularly the occipital lobe and part of the parietal lobe, is devoted to vision. Brains have no control areas or synchronisers; they rely on dialogue between circuits among which the work-load is democratically divided. Information from the retina is sent to different areas in the primary visual cortex so that form, motion and colour are recognised separately. Each primary visual area sends signals to other parts of the brain. The secondary visual cortex (in the parietal lobe) integrates them into meaningful messages. Other regions of the cortex correlate them with simultaneous inputs from other senses; and in the limbic area and other regions below the upper cortex, emotional responses are organised. Subjectively, the experience of seeing something or someone and responding emotionally is unitary. In fact, diverse brain areas are involved, each performing a distinct task.

The human brain is apparently hard-wired to attend and respond to the sight of human faces. As a result, it is able to learn very early in life to respond to particular faces, especially those of adult carers. The memories of particular faces seem to be stored towards the front of the brain in the temporal lobes. However, actual recognition of a face within the field of vision involves a region called the fusiform gyrus, where the occipital and temporal lobes meet. Therefore, matching current visual information to memory requires communication between quite separate brain areas.

From an evolutionary point of view, our skill in recognising and distinguishing faces probably served two main purposes. First, it facilitated social bonding, particularly between child and parent - a prerequisite for the prolonged learning period that became more and more necessary for survival as hominid evolution progressed. Second, it afforded a channel for communicating emotions. Communication of emotions by facial expression (which rapidly incites particular forms of behaviour and can be of great survival value) is not confined to humans. Its prevalence in mammals was the subject of Darwin's final book and has since been a major topic in ethology. However, it is particularly refined and developed in humans.

Like visual perception itself, our emotional responses to facial expression involve several different brain areas. An expression of disgust activates a region of the midbrain called the inula, which lies near the taste centres. Stimulation of the inula depresses appetite for food. The advantage is obvious: if you eat something nasty then your face expresses disgust, and no one who sees your expression feels hungry any more. Expressions of fear and anger, on the other hand, activate the amygdala, an almond-sized area that lies under the temporal lobes. Anger and fear in a tone of voice also stimulate the amygdala, which then sends signals to a region near the midbrain aqueduct called the preaqueductal grey. This initiates defensive body postures and movements, increases the pain threshold and instigates an adrenaline surge – all appropriate responses to a physical threat. However, the amygdala is not involved in recognising other emotions, or in integrating other visual and auditory signals.

Just as human brains are hard-wired for facial recognition, so they are hard-wired for acquiring language. Near the front of one temporal lobe - usually the left - is a region (Broca's area) that is necessary for speech production. Further back in the temporal lobe is a region (Wernicke's area) necessary for interpreting spoken language. Near Wernicke's area are regions involved in the processing and retrieval of verbal memories. Damage to these areas impairs recollection of what has been heard or read, but has no effect on language-using skills *per se* (or on "intelligence").

Language sounds and the meanings of words are stored - separately from one another - in parts of the left temporal cortex near the secondary auditory area. Some brain injuries damage the phonological stores without harming the lexical ones, and vice-versa. Different categories of words (parts of speech, types of noun and adverb, etc.) are processed separately in the lexical stores. Signals from these enter brain areas around the Sylvian fissure (between the temporal and frontal lobes) that are active in language production and particularly in the recall of nouns. The fine division of labour between these different parts of the brain reflects different facets of language (sounds, words etc.) and the distinct processes of recognition, interpretation, recall and production.

As with other characteristically human attainments, most of our language skills are learned rather than innate. The brain of the newborn infant is constructed to be good at language acquisition, but of course it has not yet *acquired* language. Its ability to do so seems to depend on recognition of repeated sound patterns. When strings of nonsense syllables are read monotonously to young infants (so that changes of intonation have no effect), the infants very quickly pay attention to two- and three-syllable strings that recur. The infant brain seems to be able to compute the probabilities of sound sequences and respond accordingly, treating repeated patterns as significant.

Language involves many other skills than sound recognition. It is unlikely that all these skills evolved simultaneously. Martin Nowak and his colleagues used evolutionary game theory to construct a three-step model for language evolution. First, the vocalisations of the common ancestor of humans and chimpanzees developed into a more elaborate repertoire of sounds, each sound associated with a specific object. According to Nowak's model, sound-object associations were likely to arise in a highly social species with an elaborate life-style, because the resulting ability to communicate information benefited both "speaker" and listener. Presumably the vocal apparatus became more elaborate at this time. Perhaps it happened during the proposed semi-aquatic phase of human evolution.

However, the increasing complexity of hominid life outpaced the increase of vocalisation repertoire, which has an upper limit; in all human languages, the total number of distinct phonemes is quite small. A second stage of language evolution ensued; brains that could combine sounds into words became advantageous. A repertoire of *words* rather than single sounds allowed a virtually unlimited number of objects to be represented distinctly. The third step, syntax, enabled individuals, actions and relationships to be specified uniquely. This became essential when each individual in the community had to meet a complicated range of expectations. Any reasonably advanced social learning must have required syntax. It is generally supposed that language complete with syntax did not exist before *H sapiens*, but the evolutionary conditions for syntax according to the Nowak model might have obtained earlier. The skull contours of *erectus* and even *habilis* suggest advanced temporal lobe development and therefore, conceivably, language development.

The most distinctive feature of *sapiens*, perhaps related to the increased size of the frontal lobe, seems to be the capacity for abstract thought and abstract associations. A fully developed capacity for language might have been a prerequisite for abstraction. Abstraction was in turn a prerequisite for symbolic representation, which is why the 50,000 year old Australian rock art alluded to earlier was surely the work of *H sapiens*.

Africa today is home to a large number of language families. In other parts of the world, individual language families such as Semitic and Indo-European tend to cover bigger land areas. Does this pattern reflect the migrations of *H sapiens*, originating in Africa, or the migrations of *H erectus*; or neither? The finding that human genetic variance within Africa exceeds that in the rest of the world indicates the African origin of our species. In the same way, the finding that Africa is home to the widest variety of language families might indicate that human language capacity, including syntax, was complete before our ancestors migrated out of the continent.

Minds

A vast amount has been written about the relationship between mind and brain and we cannot review the topic here. But one question is particularly interesting: why, when our understanding of brain function is burgeoning thanks to advances in neurobiology, does so much debate about minds continue? The following statements (and others of similar kind) are affirmed by some writers and denied by others:

- brain is objective while mind is subjective;
- brain is material while mind is apparently immaterial;
- brain is not "about" anything but mind is - mental processes *represent* things.

These assertions suggest that mind cannot be explained in terms of brain. It is very difficult to free oneself of the intuition of "dualism", i.e. that our "selves", our minds, are distinct from but somehow inhabit our bodies.

Some interesting experiments conducted during the 1970s and 1980s demonstrate that when we form an intention to do something, e.g. to perform a simple act such as bending a finger, changes take place in the brain hundreds of milliseconds *before* we make the "conscious intent". In other words, the relevant events in the brain precede the mental event – the "act of will". Thus, minds are in a real (though still obscure) sense *caused* by brains.

There is plenty of evidence to support this claim. Sensations such as pain in a specific part of the body can be evoked merely by stimulating the right neurones in the cerebral cortex. Specific memories can be evoked in a similar way. Every mental event is not merely accompanied by a particular brain event, often involving a great many neurones, it is immediately preceded by the brain event.

However, there are good reasons for rejecting the extreme reductionist view that "mind" is not worth considering, that we should focus attention exclusively on studying the brain instead. For one thing, the advice is impossible to follow. Much of our everyday language and thought presupposes the existence of mind. To replace attitudinal terms such as "desire", "appreciate", "believe" and "dislike" with descriptions of the relevant brain events would be absurd, even if it were possible in practice. Also, to deny the existence of mind is self-contradictory; it is tantamount to declaring "I do not exist". To say that "mind" is not a useful concept is rather like saying that "life" is not a useful concept. In the first half of this book we argued that life is matter organised in an autonomous, high-order way: a self-sustaining reciprocal dependence among gene expression pattern, responses to external stimuli and internal state. Analogously, we might say that *mind is brain function organised in a self-sustaining high-order way*. Roughly speaking, mind is to brain as life is to cell. We shall explore this analogy in chapter 18.

A brain capable of intelligent behaviour continually correlates information from the rest of the body with information from the perceived environment. There is fine division of labour, but higher-order brain processes integrate the parts, creating (for instance) connected experiences by combining impressions of form, colour and movement (derived from vision) with information from other senses such as sound, and with appropriate emotions and actions. The brain must also ensure that its integrated image of the world, and the behaviour that it initiates, is consistent from moment to moment. A brain that can do all this is a valuable tool for survival. If the number of synaptic connections between sensory input and behavioural output is sufficiently great, it is also a necessary – and perhaps sufficient - condition for mind.

However, it is fair to ask whether a human brain, if it could be kept alive in isolation, receiving no sensory inputs and able to generate no behavioural outputs at all, could "cause" (or be directly associated with) a mind. The answer is probably "no", just as an isolated cell would cease to live if it were deprived of all signals from the environment and rendered incapable of specific outputs. Several well-known experiments have shown that sensory deprivation quite rapidly disorientates human subjects; in effect, they begin to lose their minds. Minds are "caused" by brains that are actively processing sensory inputs and generating outputs, not by brains in isolation.

A brain that can abstract as well as order the information it receives from the body, and can express it via (e.g.) language, thereby creates a representation, a subjective mental state. This is what the human brain does. The subjective/objective division between mind and brain might therefore be less problematic than is often claimed. Moreover, abstract association integrates the brain's representations of self and world into a continuous and consistent unitary experience. This unitary experience is our sense of self, i.e. consciousness. So consciousness might not be hopelessly resistant to biological explanation, as (for example) Cairns-Smith and Chalmers maintain - but nor is it a mere "linguistic confusion", as Dennett claims.

In short, the mind-brain relationship might be less problematic than is generally believed, so long as it is viewed from the appropriate perspective: i.e. how the human brain evolved, and what biological purpose it serves. This perspective no more denigrates mind than our characterisation of the living state denigrated life. We believe that "consciousness" is neither a linguistic confusion nor a mystery that cannot be assimilated into conventional biology.

The capacity to abstract and articulate brain processes, enabling us to make long-term predictions, is uniquely developed in modern humans. This capacity, the root of our culture, evolved under very specific circumstances. Ancestral hominids that were already capable of intelligent behaviour produced bipedal and very highly social descendants. The sheer improbability of this combination of circumstances (intelligent behaviour, high social development and bipedalism coinciding in the same species) could explain why "human intelligence" has only evolved once on Earth and is unlikely to be simulated anywhere else in the universe (chapter 15). Other animals that behave intelligently must have higher-order integration of their brain functions, but they lack our capacity for abstract representation. They no doubt have a sense of self and can detect patterns and continuity within themselves and the world around them; but without the ability to abstract or to express these abstractions, they cannot have the kind of mental lives that humans have.

At what stage in human evolution did "mind" appear? Like language, it probably emerged step by step. Evidence of material culture at a particular time in human evolution might indicate the level of "mind" at that time. If so, there is an interesting implication. If our brains are continuing to grow as our culture becomes more complex, then our *minds* are still evolving. This inference will not appeal to evolutionary psychologists.

In chapter 18 we shall expand these arguments and clarify them.

Chapter 18

CELLS, BRAINS AND COMPUTERS: TOWARDS A CHARACTERISATION OF MIND

In chapter 10 we offered a characterisation of "life". Now we shall suggest an analogous characterisation of "mind". In chapter 17 we claimed that "mind is to brain as life is to cell", that the relationship between livingness and the cell is analogous to the relationship between mind and the brain. Using this analogy, we shall try to throw some light on the mind-brain problem in a way that might prove acceptable to biologists and to others.

"Brain state"

In chapter 6 we defined the internal state of a cell: a set of reciprocal dependences involving structure, metabolism and transport. Internal state became central part of our characterisation of "livingness". Similarly, "brain state" is an important part of our tentative characterisation of "mind".

Essentially, the *structure* of the human brain consists of neurones connected via synapses to form pathways or circuits[50]. Brain *function* consists of the activities of these circuits: ordered sequences of action potentials, neurotransmitter release events, and postsynaptic responses. Function obviously depends on structure. However, as we showed in chapter 16, structure also depends on function. Neuronal activities alter the strengths of synapses and they also forge new growth in axon termini and the formation of new connections, i.e. new circuits.

---

[50] Anyone with a basic knowledge of anatomy knows this to be an oversimplification. Apart from neurones, brains also contain large numbers of glial cells, whose functions include regulation of neurotransmitter levels. Blood vessels abound. Also, there are spaces filled with cerebrospinal fluid, which act as shock absorbers as well as maintainers of fluid, electrolyte and nutrient balance. Protective fibrous sheets surround the whole organ. When we speak of "structure" in the text we mean only those aspects of structure that are *directly* relevant to brain function.

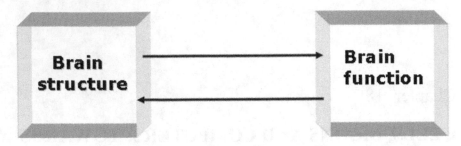

**Fig. 18-1: structure and function in the brain are reciprocally
dependent, just as structure and metabolism in the cell are reciprocally
dependent.**

This diagram recalls the reciprocal dependence between cell structure
and metabolism (i.e. cell function) discussed in chapter 4.

The brain is self-regulating. To take a simple example: when you bend
your elbow the biceps muscle contracts and the triceps relaxes. When you
straighten it again the opposite happens: the biceps relaxes and the triceps
contracts. The part of the brain responsible for these movements contains
reciprocal control systems. When the nerves to the biceps fire, the nerves to
the triceps are inhibited, and vice-versa. This control depends crucially on
the construction of the nerve pathways. For example, one set of axon
terminal branches activates an excitatory nerve, one an inhibitory nerve[51]. It
also depends on pathway function: which neurones become active, which
transmitters are released at which synapses. In turn, the control mechanism
ensures that some neurones and synapses are active and others are not.
Since brain function affects brain structure, the control processes also
indirectly affect brain structure.

Some of the consequences of this are familiar from everyday experience.
Skills such as riding a bicycle or typing have to be learned. During such
learning, the parts of the brain that control the relevant muscles are modified.
Their structural organisation is changed and therefore so is their function.
Structure, function and control in neural circuits all depend on one another.

---

[51] The motor nerves (the ones that actually cause muscle contraction) respond to two different
neurotransmitters, acetylcholine (which activates) and gamma-aminobutyrate (which
inhibits). An instruction to contract the biceps releases acetylcholine on to the biceps
nerve and gamma-aminobutyrate on to the triceps nerve. Some poisons such as strychnine
cause convulsions by interfering with the gamma-aminobutyrate inhibition, causing both
opposing sets of muscles to contract at one. The effect is to tear muscles and tendons,
break bones and cause exhaustion. Homeostasis within the brain is essential for survival.

There are more intricate examples, but they all depend on the principle that the probability of an action potential in a particular neurone at a particular time depends on activities in other parts of the brain and on the connections of these parts to the neurone. We define the "brain state" at any instant by the following diagram:-

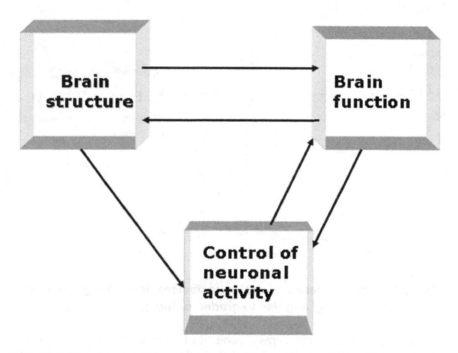

Fig. 18-2: "brain state" is roughly analogous to the "internal state" of a cell as defined in chapter 6.

"Brain state" bears some comparison to the internal state of a cell (chapter 6), although cellular transport has no obvious counterpart in the brain. Like a cell's internal state, brain state changes from moment to moment. The structures and functions of the numerous circuits, and the control process operating in them, are never constant. Since "brains cause minds", it follows that the workings of the mind are underpinned by an ever-shifting pattern of activities and an ever-changing set of connections among the 10^{15} or so synapses in the human brain.

Neurone function compared to gene expression
The following diagram shows a simple abstract model of an analogue computing device. A variety of inputs (I) feed into an integrator (C). Some inputs are positive (+), some are negative (-). Some act indirectly, blocking

or augmenting other inputs.   The integrator (C) sums the inputs from moment to moment and instructs an effector (E) to vary the output (O) accordingly.   The output varies from zero to a maximum value that depends on the detailed construction and the function of the device.

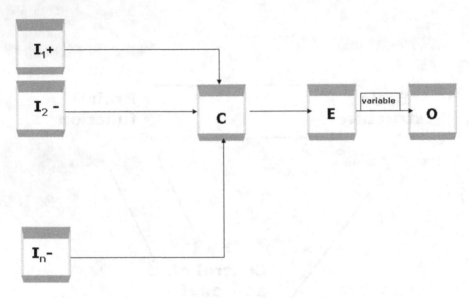

Fig. 18-3: cells and genes as analogue devices, integrating inputs and giving rise to graded outputs.

Suppose the device represents a gene. In this case, the output (O) is the transcription rate (the number of messenger RNA copies made per second). It is continuously variable from zero to a maximum. The inputs (I) are the enhancers occupied by transcription factors; most are positive, but some are negative. The integrator (C) is the initiation complex, the activity of which depends on the sum of the inputs. The effector (E) is RNA polymerase II. Thus, the diagram is an abstraction of the control of gene expression as described in chapter 8.

Now, suppose instead that the device represents a neurone. In this case, the output (O) is the rate of firing (the number of action potentials per second). It is continuously variable from zero to a maximum. The inputs (I) are the postsynaptic potentials at the dendrites and on the cell body; some of these are excitatory and some are inhibitory. The integrator (C) is the cell body, whose grand postsynaptic potential represents the sum of the inputs. The effector (E) is the axon hillock, where action potentials are initiated. Thus, the diagram is an abstraction of neurone function as described in chapter 16.

This abstract model can be elaborated. For example, the integrator (C in the diagram) can be modulated, making it more or less sensitive to positive or negative inputs, or more or less able to activate the effector. In genes, chemical modification of the initiation complex proteins makes the initiation complex more or less inclined to launch the polymerase. In neurones, calcium channels in the cell body can make the axon hillock leak potassium ions and become less inclined to launch action potentials.

The purpose of gene transcription is to make the cell's proteins. Messenger RNAs are translated and proteins are produced. The proteins are responsible for the functional organisation of the cell as a whole. Analogously: the purpose of action potentials is to activate and inhibit synapses. Neurotransmitters are released and postsynaptic receptors are occupied. Synapses are responsible for the functional organisation of the brain as a whole.

To summarise:-

Component/function	Gene expression	Neuronal activity
Unit	Gene	Neurone
Inputs	Transcription factors; Enhancer regions of DNA	Neurotransmitters; Receptors on dendrites and cell body
Integrator	Initiation complex	Cell body
Effector	RNA polymerase II	Axon hillock
Output	Transcription	Action potential (impulse)
Output variation	Transcription rate	Firing rate
Integrator modulation	Chemical modulation of initiation complex proteins	Calcium channels
Consequence of unit activity	Proteins made	Synapses activated/inhibited

The "consequences of unit activity" include inputs to other units (genes or neurones). Transcription factors are proteins, so the expression of one gene (encoding a transcription factor) can affect the expression of others. Action potentials release neurotransmitters from axon termini, so the activity of one neurone can affect the activities of others. In the diagram, the "O" of one unit is an "I" for one or more others.

Neurone and brain state compared to gene and internal state

The analogy between neurone activity and gene expression is therefore striking. Earlier in this chapter we compared the brain state with a cell's internal state. So it is tempting to infer that *neurone activity is to brain state as gene expression is to internal state*. However, this would be slightly misleading. In chapters 8 and 9 we discussed the significance of the time delay between gene expression and change of internal state. Neuronal activity affects brain state *immediately and directly*. Genes (DNA) are not part of the internal state, which depends on proteins. Neurones *are* part of the brain state, which depends on neurones and synapses. However, the comparison can still be made:-

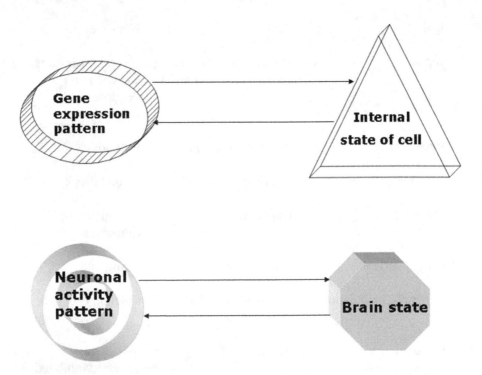

Fig. 18-4: Further development of the analogy between cell and brain.

We should emphasise the distinction between the left and right hand sides of the lower diagram. The activities of individual neurones can be detected by inserting microelectrodes into the brain. Brain state is indicated by EEG recordings, PET scans and other techniques for examining the activity of the brain as a whole. Brain states differ between sleeping and waking, but the activity of an individual neurone might be the same.

Stimuli and signalling pathways
In chapter 9 we introduced the cell's responses to external stimuli. This was the third apex of the internal-state/gene-expression-pattern/stimulus-response triangle by which we characterised "livingness". Now we shall introduce the brain's responses to external stimuli. This principle forms the third apex of the brain-state/neuronal-activity-pattern/stimulus-response triangle by which we might characterise "mind".

Just as a cell interacts with its environment via receptors on the surface membrane, so the brain interacts with its environment through specialised sensors. (The "environment" of the brain includes the rest of the body as well as the outside world.) So now we can extend our analogy between cell and brain:-

- The effects of stimuli are subject to adaptation. Excessive or over-prolonged inputs cease to elicit any marked response from the cell/brain.
- The immediate effect terminates when the stimulus ceases.
- A single stimulus usually affects several widely-separated functions within the cell/brain.
- Any function within the cell/brain can be modified by several different types of stimulus, which might or might not be experienced simultaneously.
- The effect of the stimulus is transferred from the receptor/sensor to the point(s) of action within the cell/brain by a more or less long sequence of events, a "signalling pathway".
- There is extensive cross-talk among signalling pathways.

Sensors such as light-sensitive cells in the retina, vibration-sensitive cells in the inner ear and the pressure-sensitive cells in major arteries all activate or inhibit neurones, sending impulses to the brain. The mechanism by which an external stimulus is converted to a change in neurone firing rate can be complicated[52] but the effect is simple: there is a change in the frequency of action potentials in neurones connecting the sensor to the brain.

---

[52] For example, consider a rod cell in the retina. When the cell is dark-adapted, its membrane sodium channels are jammed open by a "molecular wedge" (a cyclic nucleotide). In this state, the cell is active. It releases inhibitory neurotransmitters and prevents signals travelling to the brain. When light strikes the cell, a membrane

Cross-talk among neuronal pathways can take place at all levels between the sensor and the cerebral cortex. A signal to which the brain is primed to respond, perhaps by memory (facilitated synapses), can attenuate other simultaneous signals. For instance, you can switch attention from one nearby conversation to another during a noisy party. A woman can sleep through a thunderstorm but wake up when her baby cries. These examples involve pathway cross-talk in the higher parts of the brain. Near the sensor itself, processes such as *lateral inhibition* sharpen the focus of a signal pathway. Lateral inhibition works roughly as follows. Suppose an external stimulus activates five sensor cells: A and E rather weakly, B and D moderately and C strongly. At the first synapse after the sensory surface, each cell stimulates its own postsynaptic neurone but inhibits those on either side. Thus, although five sensory cells are activated, only the neurone from C carries information to the brain.

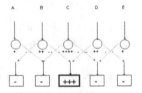

Fig. 18-5: schematic illustration of lateral inhibition.

Thanks to lateral inhibition, we can see sharp edges even though several receptor cells in the retina might be stimulated. We can hear pure tones in music even though several neighbouring hair-cells in the inner ear all vibrate to a greater or lesser extent.

---

protein (rhodopsin) changes its shape and activates an enzyme that removes the "molecular wedge". The sodium channels then close, the cell is de-activated, and no inhibitory neurotransmitters are released; so a signal is sent to the brain. This mechanism seems Byzantine but it is highly efficient, because the rhodopsin shape change also leads to a slow manufacture of "wedge" molecules, enabling the cell to *adapt* to continued light irradiation.

Mind and the analogy with the living state

Inputs to the brain from sensors alter the activities of particular neurones. They also alter the "brain state". But the ability of the sensors to deliver information to the brain depends *on* the brain state. An obvious example is the difference between a sleeping person and an alert one. Also, "control" neurones can alter the responses of the sensors. Processes similar to lateral inhibition can enable particular pathways in the brain to facilitate or inhibit specific sensory inputs.

In short, we can summarise the brain's activities by a diagram similar to the one that summarised our characterisation of the living state (chapter 10).

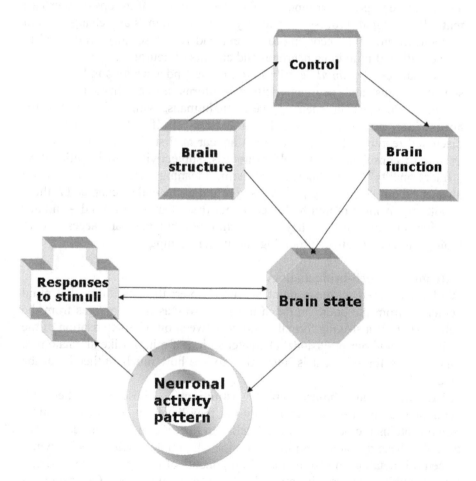

Fig. 18-6: "Mind is to brain as life is to cell". Compare Fig. 10-1.

There are a few differences of detail:-

- Nothing in the brain corresponds to "transport" in the cell's internal state.
- The arrow connecting "structure" to "control" is one-way.
- The times (t_1, t_2 and t_3) that appeared on the arrows in the living-state diagram in chapter 10 are missing here.

This diagram could be taken to characterise the mind of any animal capable of intelligent behaviour, as defined in chapter 16. At any moment, stimuli from the body and from the surroundings are activating sensors, so information is being sent to the brain. Which sensory information is processed depends on the brain state as a whole (whether the animal is alert and attentive to particular inputs) and on the activities of groups of neurones that affect signal processing pathways. The brain state changes from moment to moment, according to current and recent sensory inputs and to the activities of individual neurones and circuits of neurones.

We suggest that *mind* describes the interdependences among brain state, sensory inputs and neuronal activity patterns at any instant, and the continuity of these interdependences. In humans, with their capacity for abstraction and language, mind can express itself. Hence the much-discussed features of mind that do not appear at first sight to be features of the brain: consciousness or self-awareness, subjectivity and intentionality. ("Intentionality" is a philosophical term, meaning that our thoughts are always "about" something.) As we suggested at the end of chapter 17, there is nothing mystical or non-biological about these features of mind – and nor are they empty words. They are meaningful, but they are accessible to biological (specifically neurobiological) understanding.

Extending the cell-brain analogy

We have covered the main point of this chapter: the possibility that mind "emerges from the brain" in much the same way as life "emerges from the cell". There is a striking formal analogy between our characterisation of the living state and our provisional characterisation of mind. Like all analogies, this one has limits, but it is interesting to see how much further it can be extended.

1. "Luxury" and "housekeeping" functions. These are rather old-fashioned terms in molecular biology but they have not lost their meaning. Some proteins are necessary for all cells, so they are made in all cells. We describe these as "housekeeping" proteins. Examples include the enzymes of central metabolism and the main components of the cytoskeleton. Genes for these ubiquitous proteins are called "housekeeping" genes. Other proteins are only necessary, at least in significant quantities, in particular differentiated cells, i.e. cells that have become dedicated to specialist functions at the expense of their capacity to divide. Genes for these specialist proteins are called "luxury" genes.

Some parts of the brain are necessary for overseeing basic physiological functions, e.g. respiratory muscle contractions and the beating of the heart. No vertebrate could survive without these functions. Other parts of the human brain are devoted to specifically human functions such as facial recognition, communication of emotions, language and abstract thought. The basic-physiology parts of the brain can be regarded as analogous to "housekeeping" genes, and the higher-function parts to "luxury" genes.

2. Redundancy. There is a good deal of redundancy in both cell and brain. Many signalling pathway components in cells are the products of genes that have been duplicated and then modified during the course of evolution[53]. In the brain, "back-up" systems can be brought into play if an area is damaged; and sensory inputs are parcelled into blocks of information that are sent to disparate areas. These are examples of redundancy. The human visual system has at least six distinct and apparently largely independent processing pathways. Apparently this is the result of duplication during embryonic brain development.

Redundancy in both cell and brain provides emergency back-up mechanisms in the event of system failure. It also affords more possibilities for cross-talk among different signalling pathways. And as Kauffman and others have observed, redundancy seems to be a prerequisite for robustness in complex systems.

3. Stress responses. When external conditions exceed the limits of normal cell functioning, "heat-shock" genes are expressed. The resultant proteins shut off almost all luxury functions. Just enough housekeeping functions are sustained to keep the cell alive. The heat-shock proteins bind to regulators of luxury function. They release them again when conditions become less hostile and normal cell activity can be resumed.

When conditions exceed the capacity of the brain to tolerate them, individuals suffer clinical depression, or in some cases catalepsy. Many higher brain functions are down-regulated, though housekeeping activities remain intact. Stress hormones are released, ensuring a continuing glucose supply. This is necessary (though it is unlikely to be sufficient) for restoring normal brain function.

---

[53] There are numerous examples. A prominent one is protein kinase C, which is essential for transducing many extracellular signals. Protein kinase C exists in many different isoforms, the products of a much-duplicated and modified gene.

There is therefore some parallelism between the effects of "stress" on the cell and the effects of "stress" on the brain.

Are humans special after all?

In chapter 15 we suggested that "human intelligence" is an evolutionary freak, so improbable that nothing like it has arisen elsewhere else in the universe. We repeated this claim in chapter 17. We have also argued that although any animal capable of intelligent behaviour can be said to have a mind, humans are unique in being able to express and articulate the contents of their minds.

This conclusion is scarcely original or startling. All we have done in the past three chapters is to relate mind to brain and to evolution in a rather novel (though in no way remarkable) way. But in conjunction with our claim for human uniqueness, perhaps this does lead to a surprising inference. If humans are unique in the nature and articulation of their minds, and if they have no counterpart anywhere else in the universe, then the capacity to abstract and express mental processes is *absolutely* unique to us. If so, the evolutionary emergence of our species has added something entirely new to the universe. This is an optimistic inference, very much at odds with the existential pessimism that seems to have become part of the legacy of science.

Galileo kicked the immovable Earth into orbit around the sun. To his more conservative contemporaries, his temerity seemed to denigrate humankind. Since Galileo's day the Earth, and the sun around which it rotates, and even the galaxy of which the sun is one tiny part, have grown more and more insignificant in relation to the universe as a whole. Many people find this depressing. The Earth is utterly insignificant, they say, so what is the point of our being here?

Darwin kicked the Aristotelian ladder of nature from under the living world. In the process, he revealed that the human species is a late-comer among countless millions of species of organisms, inherently no more special than any other, and guaranteed to share the ultimate fate of all – extinction. Many people find this even more depressing than the apparent insignificance of our planet.

What Galileo, Darwin and their successors told us is perfectly correct. Nevertheless, if our argument in the past four chapters is valid, our species is unique and special. If we are right, then only on our apparently insignificant planet, and in our apparently insignificant species, is there such a thing as a self-expressing mind. Without us, the cosmos would have no mind that could record its existence.

What we, as a species, have done with this cosmically unique innovation has not always been interesting or commendable. However, some of it has. What we shall do with it in the future is another matter.

In the meantime, we can enjoy our unique capability. Sitting in the quiet woodland at sunrise, marvelling at the primrose and the oak-tree, the beetle and the weasel, we can reflect on the fact that nowhere else in the universe is there an entity with this capacity for marvelling, or for inquiring and understanding what "life" is; or for inquiring what constitutes the "capacity for marvelling".

GLOSSARY AND PRONUNCIATION GUIDE

NOTE: cross-references to other parts of this guide are indicated by the use of UPPER CASE letters. Where pronunciation is given, the stressed syllable is indicated by the mark'.

Action potential: Brief electric impulse that travels along the AXON in a NEURONE.

Amino acid (pron. A-my'-no acid): A building block (monomer unit) of a PROTEIN molecule.

Amygdala (pron. A-mig'-duh-la): Small piece of the brain below the temporal lobe responsible for emotions of anger and fear. (Greek: *amygdale* = almond.)

Anabolism (pron. A-nab'-ole-ism): Any process by which biological molecules are manufactured in cells. (Greek: *an* = up, + METABOLISM.)

Archaea (pron. Ar-key'-a): PROKARYOTE group that includes the inhabitants of extreme environments. (Greek: *archaios* = ancient, *arche* = beginning.)

ATP = adenosine triphosphate: Molecule crucially involved in energy-requiring and energy-mobilising reactions in cells. (Greek: *aden* = gland.)

Axon: Extension of a NEURONE that carries ACTION POTENTIALS. (Greek: *axon* = axis.)

Base: Part of a building block (monomer unit) of a nucleic acid; the other parts are phosphate and a type of sugar. Sequences of bases in DNA constitute genetic information. (Note: in chemistry, "base" has a different and more general definition, not relevant to this book.)

Biosphere: Parts of the Earth and atmosphere in which living things are found. (Greek: *bios* = life.)

Carbonaceous chondrite (pron. car-bon-ay'-shuss kond'-right): A meteorite containing rounded granules with organic molecules. (Greek: *chondros* = grain, granule.)

Catabolism: Any process by which biological molecules are broken down to release energy. (Greek: *cata* = down, + METABOLISM.)

Catalysis (pron. Cat-a'-liss-iss): Chemical change brought about by a substance (in biology, an ENZYME) that is itself unchanged. (Greek: *cata* = down, *lysis* = dissolution.)

Cell membrane: The ultra-thin barrier that divides the cell from its environment.

Cell wall: The tough shell around a prokaryote or a plant or fungal cell.

Chloroplast (pron. Kloa'-row-plast): A green body inside a plant cell that is responsible for photosynthesis. (Greek: *chloros* = pale green, *plasma* = body.)

Coelenterazine (pron. See-lent'-er-a-zeen'): A molecule that binds oxygen, derived from marine organisms such as coelenterates. (Greek: *koilos* = hollow, *enteron* = intestine.)

Cyanobacteria (pron. Sigh-an'-no-bacteria): Prokaryotes containing chlorophyll, the green pigment essential for photosynthesis. (Greek: *kyanos* = blue.)

Cytoskeleton (pron. Sigh'-toe-skeleton): The fibrous structures in cells responsible for shape, movement and some transport processes. (Greek: *kytos* = vessel, hollow.)

Dendrite: Branching projection of a NEURONE. (Greek *dendron* = tree.)

DNA: Deoxyribonucleic acid. The substance of which genes are made.

Ecosystem: A community of organisms, their environment and the interactions among them. (Greek: *oikos* = house.)

Ediacara fauna: The earliest known multicellular organisms, dating from the late Precambrian. (Named after the Ediacara Hills, Flinders Range, South Australia.)

Endocytosis (pron. End'-oh-sigh-toe'-siss): Uptake of material into cells by pinching off of membrane vesicles. (Greek: *endo* = within, *kytos* = vessel, hollow.)

Enhancer: Region of DNA that binds TRANSCRIPTION FACTORS, modifying the rate of TRANSCRIPTION of one or more genes.

Enzyme: A biological CATALYST. Most enzymes are proteins; a few are RNA.

Eukaryote (pron. You-car'-ry-oat): An organism consisting of one or more cells, each with its DNA packaged in a separate compartment, the NUCLEUS. (Greek: *eu* = well (formed), *karyon* = kernel.)

Exon: Segment of a gene that codes for part of a protein. (Latin: *ex* = out of, + Greek neuter suffix *-on*.)

Exon shuffling: A DNA rearrangement in which an EXON of one gene is inserted into a different gene.

Forebrain: The most recently evolved part of the brain, including the cerebral hemispheres.

Frontal lobe: Part of the cerebral hemispheres behind the forehead. In humans, associated with (among other things) the capacity for abstract thought.

Gene duplication: A DNA rearrangement in which two or more copies of the same gene are inserted into the DNA. (Greek: *-genes* = born.)

Gene pool: The stock of genes and gene variants found in an inbreeding POPULATION.

Genome: The total stock of genes in an individual (assumed to be a typical member of the species.)

Hindbrain: Evolutionarily oldest part of the vertebrate brain. Includes the cerebellum, which is involved in the control of bodily movements, and the medulla oblongata, a part of the brain stem involved in controlling respiration and heart rate.

Hydrocarbon: A chemical substance consisting only of the elements hydrogen and oxygen.

Immediate-early genes: Genes that are activated first during a developmental process. In turn they activate many other genes, directly or indirectly.

Initiation complex: Assembly of proteins that enables RNA POLY-MERASE II to begin TRANSCRIPTION at the start of a gene.

Intron: A non-coding segment of a gene; opposite of EXON.

Kaolinite: A flaky aluminium silicate mineral produced by the breakdown of feldspar.

Lysosome (pron. Lie'-so-soam): A membrane-bound particle containing enzymes necessary for intracellular digestion. (Greek: *lysis* = dissolution, *soma* = body.)

Metabolic pathway: A sequence of chemical reactions in a cell that converts one sort of molecule into another.

Metabolon: An assembly of enzymes needed for (part of) a metabolic pathway.

Midbrain: Part of the brain dominant in reptiles. In mammals, links the forebrain and hindbrain and contains areas involved in emotional responses, pain, etc.

Mitochondria (pron. My'-toe-kon'-dree-a): The energy-producing bodies in EUKARYOTIC cells. (Singular = **mitochondrion**. Greek: *mitos* = thread, *chondros* = granule.)

Montmorillonite (pron. Mont-mur-ill'-o-night): A clay mineral chemically similar to KAOLINITE; a constituent of Fuller's Earth. (Named after Montmorillon, France.)

Mutation: Genetic change. (Latin: *mutare* = to change.)

Neurone: A nerve cell. (Greek: *neurone* = nerve.)

Neurotransmitter: A chemical released from an AXON terminus that conveys the nerve impulse across a SYNAPSE.

Nucleic acid: A polymer of which the building blocks (monomer units) are molecules consisting of a sugar, a phosphate and one of four bases: A, G, C, T in DNA; A, G, C, U in RNA.

Nucleus: The part of an EUKARYOTIC cell that contains most of the DNA (genes).

Occipital lobe: Part of the cerebral cortex at the back of the head.

Parietal lobe: Part of the cerebral cortex at the top of the head. (Latin: *paries* = wall.)

Polymer: A molecule made by joining many smaller molecules (monomer units or "building blocks") together. (Greek: *poly* = many, *meroi* = parts.)

Polymerase: An enzyme that joins "building block" molecules together to make a biological polymer.

Population: All the members of a species that are capable (for geographical and other reasons) of breeding with one another.

Prokaryote (pron. Pro-car'-ry-oat): A single-celled organism that does not have its DNA packaged in a separate nucleus. (Greek: *pro* = before + *karyon* = kernel.)

Protein: A polymer of AMINO ACIDS. Proteins are responsible for all the structural, functional and informational features of a cell.

Protist: A unicellular EUKARYOTE. (Greek *protistos* = the very first.)

Pseudogene: A degenerate copy of a gene incapable of being transcribed.

Receptor: Structure on a cell (surface) that binds specifically to a stimulus (signal) molecule and initiates a response in the cell.

Resting potential: Electrical potential between the inside and the outside of a NEURONE or other cell.

Retrotransposon (pron. Ret'-row-trans-po'-zon): A piece of DNA containing retroviral and other genes that can move from place to place in the genome.

Ribosome (pron. Righ'-bow-soam): An intracellular machine for making proteins according to the instructions in messenger RNAs.

RNA: Ribonucleic acid. Includes messengers that carry the instructions in a gene to the protein-making machinery (RIBOSOMES).

RNA polymerase: An ENZYME for making RNA (i.e. for TRANS-CRIBING it from a sequence in DNA).

Signalling pathway: A sequence of reactions in a cell that is initiated by the binding of a stimulus molecule to a receptor. A signalling pathway can lead to various changes in cell structure and metabolism and in gene expression.

Simple-sequence DNA: A short segment of DNA (3-10 bases) that is repeated hundreds or even thousands of times in the GENOME.

Spirochaete (pron. Spy'-row-keet): A type of spirally coiled bacterium that swims, usually by making "wriggling" movements. (Greek: *speira* = coil, *chaite* = hair.)

Symbiosis (pron. Sim-buy-owe'-siss): Mutual dependence between organisms of two or more different species. (Greek: *syn* = together, *bios* = life.)

Synapse: The tiny gap between one NEURONE and another. (Greek: *syn* = together, *haptein* = to fasten.)

Taxonomy: Biological classification. (Greek: *taxis* = order, *nomia* = distribution.)

Temporal lobe: Lobe on each side of the cerebral cortex of the brain. (Latin *tempus* = temple.)

Terminal arborisation: The branching of an AXON that allows it to form SYNAPSES with many other NEURONES. (Latin: *arbor* = tree.)

Transcription: The process by which a messenger RNA copy of a gene is made.

Transcription factor: A PROTEIN that binds to an ENHANCER region of DNA and alters the rate of TRANSCRIPTION of one or more genes.

Transposon: A piece of DNA that can move to a different position on a chromosome or to a different chromosome, altering the cell's genetic makeup.

Vacuole: A small fluid-filled cavity in a cell. (French: little vacuum.)

Vesicle: A roughly spherical structure inside a cell caused by the pinching off of a small piece of membrane. (Latin: *vesica* = bladder, blister.)

FURTHER READING

Chapters 1-10

De Duve, C (1984) *A Guided Tour of the Living Cell*, Scientific American Books, New York.

Goodsell, D S (1992) "A look inside the living cell," *American Scientist*, Sep-Oct, 457-465.

Maturana, H R and Varela, F J (1987) *The Tree of Knowledge*, New Science Library, Boston, Massachusetts.

Maynard Smith, J (1986) *The Problems of Life,* Oxford University Press, Oxford.

Smith, C U M (1976) *The Problem of Life*, Wiley, New York.

Thomas, L (1975) *The Lives of a Cell*, Viking, New York.

Bray, D (1992) *Cell Movements*, Garland, New York and London.

Harold, F M (1986) *The Vital Force: a Study of Bioenergetics*, W H Freeman, New York.

Harold, F M (2001) *The Way of the Cell*, Oxford University Press, Oxford.

Kauffman, S A (1995) *At Home in the Universe: The Search for Laws of Self-Organisation and Complexity*, Oxford University Press/Viking, London.

Lewin, R (1992) *Complexity: Life on the Edge of Chaos*, Macmillan, New York.

Pirie, N W (1938) "The meaninglessness of the terms 'life' and 'living'," in Needham, J and Green, D (eds) (1938) *Perspectives in Biochemistry*, Cambridge University Press, London.

Preston, T M, King, C A and Hyams, J S (1990) *The Cytoskeleton and Cell Motility*, Blackie and Son, Glasgow.

Rosen, R (1991) *Life Itself: A Comprehensive Inquiry into the Nature, Origin and Fabrication of Life*, Columbia University Press, New York.

Wolpert, L (1991) *The Triumph of the Embryo*, OUP, New York.

Chapter 11

Benner, H A (ed) (1988) *Redesigning the Molecules of Life*, Springer Verlag, Berlin.

Berg, D E and Howe, M M (eds) (1989) *Mobile DNA*, American Society for Microbiology.

Nei, M and Koehn, R K (eds) (1983) *Evolution of Genes and Proteins*, Sinauer Associates, Sunderland, Massachusetts.

Pennisi, E and Roush, W (1997) "Developing a new view of evolution," *Science* **277**, 34-37.

Raff, R A (1996) *The Shape of Life: Genes, Development and the Evolution of Animal Form*, University of Chicago Press, Chicago.

Terzaghi, E A, Wilkins, A S and Penny, D (eds) (1984) *Molecular Evolution: an Annotated Reader*, Jones and Bartlett, Boston, Massachusetts.

Chapter 12

Cavalier-Smith, T (1975) "The origin of nuclei and of eukaryotic cells," *Nature* **256**, 463-468.

Dawkins, R (1987) *The Blind Watchmaker: Why the Evidence of Evolution Reveals a Universe Without Design*, Norton, New York.

Elton, C S (1958) *The Ecology of Invasion by Animals and Plants*, Chapman and Hall, London.

Gold, T (1998) "The Deep Hot Biosphere," *Proc Natl Acad Sci USA* **89**, 6045-6049.

Gould, S J (1994) "The evolution of life on earth," *Scientific American* **271**, Oct, 85-91.

Grayson, A (2000) *Equinox: the Earth*, Channel 4 Books, London.

Marchant, J (2000) "Life from the skies," *New Scientist* **167**, 4-6.

Schleifer, K H and Stackebrandt, E (eds) (1985) *Evolution of Prokaryotes*, Academic Press, New York.

Chapter 13

Benton, M J (1995) "Diversification and extinction in the history of life," *Science* **268**, 52-58.

Caldeira, K and Kasting, J F (1992) "The life span of the biosphere revisited," *Nature* **360**, 721-723.

Eigen, M and Oswatitch, R W (1992) *Steps towards Life: a Perspective on Evolution*, Oxford University Press, New York.

Fortey, R (1997) *Life: an Unauthorised Biography*, HarperCollins, London.

Gould, S J (1989) *Wonderful Life: the Burgess Shale and the Nature of History*, Norton, New York.

Kauffman, S A (1993) *The Origins of Order: Self-Organization and Selection in Evolution*, Oxford University Press, New York.

Lenton, T M (1998) "Gaia and natural selection," *Nature* **394**, 439-447.

Lovelock, J (1989) *The Ages of Gaia: a Biography of our Living Earth*, Oxford University Press, Oxford.

Margulis, L (1981) *Symbiosis in Cell Evolution*, Freeman, San Francisco.

Margulis, L (1998) *The Symbiotic Planet*, Weidenfeld and Nicolson, London.

Rees, M (1997) *Before the Beginning of our Universe and Others*, Simon and Schuster, London.

Raup, D M (1991) *Extinction: Bad Genes or Bad Luck?* Norton, New York.

Schidlowski, M (1988) "A 3,800 million year isotopic record of life from carbon in sedimentary rocks," *Nature* **333**, 313-318.

Sleep, N H, Zahnle, K J, Kasting J F and Morowitz, H J (1989), "Annihilation of ecosystems by large asteroid impacts on the early Earth," *Nature* **342**, 139-142.

Szathmàry, E and Maynard Smith, J (1995) "The major evolutionary transitions," *Nature* **374**, 227-232.

Chapter 14

Baross, J A and Hoffman, S E (1985) "Submarine hydrothermal vents and associated gradient environments as sites for the origin and evolution of life," *Origins of Life* **15**, 327-345.

Cairns-Smith, A G (1985) *Seven Clues to the Origin of Life*, Cambridge University Press, Cambridge.

Crick, F C (1981) *Life Itself*, Simon and Schuster, New York.

DeDuve, C (1991) *Blueprint for a Cell: the Nature and Origin of Life*, Neil Patterson, Burlington, North Carolina.

Dyson, F (1985) *Origins of Life*, Cambridge University Press, Cambridge.

Gesteland, R F and Atkins, J F (eds.) (1993) *The RNA World*, Cold Spring Harbor Laboratory Press.

Hoyle, F and Wickramasinghe, N C (1981) *Evolution from Space*, Simon and Schuster, New York.

Mason, S F (1984) "Origins of biomolecular handedness," *Nature* **311**, 19-23.

Maynard Smith, J and Szathmáry, E (1999) *The Origins of Life*, Oxford University Press, Oxford.

Miller, S L and Orgel, L E (1973) *The Origins of Life*, Prentice Hall, Englewood Cliffs, New Jersey.

Nicolis, G and Prigogine, I (1994) *Self-Organization in Non-Equilibrium Systems*, Wiley, New York.

Russell, M J and Hall, A (1997) "The emergence of life from iron monosulphide bubbles at a submarine hydrothermal redox and pH front," *J Geolog Soc* **154**, 377-402.

Schopf, J W and Walter, M R (1983) *Earth's Earliest Biosphere*, Princeton University Press, New Jersey.

Shapiro, R (1986) *Origins: a Skeptic's Guide to the Creation of Life on Earth*, Summit, New York.

Chapter 15

Barrow, J D and Tipler, F J (1986) *The Anthropic Cosmological Principle*, Oxford University Press, Oxford.

Boss, A (1998) *Looking for Earths: the Race to Find New Solar Systems*, Wiley, New York.

Davies, P C W (1999) *The Fifth Miracle: the Search for the Origin and Meaning of Life*, Simon and Schuster, New York.

Lemonick, M D (1998) *Other Worlds: the Search for Life in the Universe*, Simon and Schuster, London.

Levay, S and Koerner, D (2000) *The Scientific Quest for Extraterrestrial Life*. Oxford University Press, London.

Orgel, L E (1992) "Molecular replication," *Nature* **358**, 203-209.

Rood, R T and Trefil, J S (1981) *Are We Alone?* Charles Scribner's Sons, New York.

Sagan, C (1994) "The search for extraterrestrial life," *Scientific American* **271**, October, 70-77.

Shostak, S (1998) *Sharing the Universe: Perspectives on Extraterrestrial Life*. Berkeley Hills Books.

Tattersall, I (1997) *Becoming Human: Evolution and Human Uniqueness*, Harcourt Brace, New York.

Trefil, J (1997) *Are We Unique?* Wiley, New York.

Chapter 16

Axelrod, R (1984) *The Evolution of Cooperation*, Basic Books, New York.

Calvin, W H (1994) "The emergence of intelligence," *Scientific American* **271**, 79-85.

Edelman, G (1987) *Neural Darwinism: the Theory of Neuronal Group Selection*, Basic Books, New York.

Hebb, D O (1949) *The Organization of Behavior*, Wiley, New York.

McGaugh, J L, Weinberger, N M and Lynch, G (eds) (1990) *Brain Organization and Memory: Cells, Systems and Circuits*, Oxford University Press, London.

Rose, S (1992) *The Making of Memory*, Transworld, London.

Sober, E and Wilson, D S (1998) *Unto Others*, Harvard University Press, Boston, Massachusetts. (Group selection supported.)

Chapters 17-18

Cairns-Smith, A.G. (1999) *Secrets of the Mind: a Tale of Discovery and Mistaken Identity*, Springer-Verlag, New York.

Calvin, W H (1994) "The evolution of intelligence," *Scientific American* **271**, October, 78-85.

Cavalli-Sforza, L L, Cavalli-Sforza, F and Thorne, S (1995) *The Great Human Diasporas: The History of Diversity and Evolution*, Addison-Wesley, Reading, Massachusetts.

Coppens, Y (1994) "East Side Story: the origin of humankind," *Scientific American* **270**, 62-69.

Cummins, D D and Allen, C (eds) (1998) *The Evolution of Mind*, Oxford University Press, Oxford.

Dennett, D C (1991) *Consciousness Explained*, Penguin, Harmondsworth.

Gardner, H (1999) *Intelligence Reframed: Multiple Intelligences for the 21st Century*, Basic Books, New York.

Horrobin, D (2001) *The Madness of Adam and Eve*, Bantam, London.

Leakey, R E and Roger, L (1992) *Origins Reconsidered: in Search of What Makes Us Human*, Little, Brown, New York.

Lieberman, P (1998) *Eve Spoke: Human Language and Human Evolution*, W W Norton, New York.

Lumsden, C and Wilson, E O (1999) *Promethean Fire: Reflections on the Origin of Mind*, Harvard University Press, Cambridge, Massachusetts.

Miller, G (2000) *The Mating Mind: How Sexual Choice Shaped the Evolution of Human Nature,* Heinemann, London.

Searle, J R (1984) *Minds, Brains and Science*, Harvard University Press Boston, Massachusetts.

Stringer, C and McKie, R (1996) *African Exodus: the Origins of Modern Humanity*, Random House, London.

Tudge, C. (1995) *The Day Before Yesterday: Five Million Years of Human History*, Jonathan Cape, London.

Wills, C (1994) *The Runaway Brain*, HarperCollins, London.

Young, J Z (1987) *Philosophy and the Brain*, Oxford University Press, Oxford.

INDEX